21st International Scientific Conference on Reshetnev Readings 2017

IOP Conference Series: Materials Science and Engineering
Volume 467

Krasnoyarsk, Russia
8-11 November 2017

ISBN: 978-1-5108-8011-5
ISSN: 1757-8981

Printed from e-media with permission by:

Curran Associates, Inc.
57 Morehouse Lane
Red Hook, NY 12571

Some format issues inherent in the e-media version may also appear in this print version.

Printed by Curran Associates, Inc. (2019)

For permission requests, please contact the Institute of Physics at the address below.

Institute of Physics
Dirac House, Temple Back
Bristol BS1 6BE UK

Phone: 44 1 17 929 7481
Fax: 44 1 17 920 0979

techtracking@iop.org

Additional copies of this publication are available from:

Curran Associates, Inc.
57 Morehouse Lane
Red Hook, NY 12571 USA
Phone: 845-758-0400
Fax: 845-758-2633
Email: curran@proceedings.com
Web: www.proceedings.com

Table of contents

Volume 467

21st International Scientific Conference Reshetnev Readings-2017

8-11 November 2017, Krasnoyarsk, Russian Federation

Accepted papers received: 16 November 2018
Published online: 7 January 2019

Preface

Papers

IOP Conf. Series: Materials Science and Engineering **467** (2019) 011001 doi:10.1088/1757-899X/467/1/011001

21st International Scientific Conference Reshetnev Readings-2017 (8–11 November 2017)

Preface to 21st edition of Reshetnev Readings Conference Proceedings.

The Russian Academician Mikhail Reshetnev (1924–1996) is one of the pioneers of the Soviet space program. His contributions were especially important to Soviet (later, Russian) satellite communication and navigation systems. As a sign of gratitude for Reshetnev's achievements, two institutions – our University and a satellite manufacturing company were named in his honor.

2017 is the 21st time that the University has organized a scientific conference in memory of Academician Reshetnev. Prior to 2017, the University was named Reshetnev Siberian State Aerospace University. After it merged with the Siberian University of Technology in 2016, its name was changed to Reshetnev Siberian State University of Science and Technology, or Reshetnev University.

2017 is, therefore, the first time that the University is holding the Conference in a new status.

While the traditions sections of the Conference with the fields of aerospace technology and engineering, computer science and telecommunication systems remain as the nucleus of the Conference, new sections, including a significantly developed section or workshop on Materials Science and Engineering have been added. The scope of the section has been broadened, adding new areas of research such as forestry products and biomaterials in engineering, solid-state physics, composite materials, polymers, and other areas related to the science of developing and producing new materials.

The Conference is traditionally held in the Siberian city of Krasnoyarsk. Over 200 participants annually take part in the Conference, making it one of the largest scientific event in the city. The Conference is multidisciplinary. However, the topics of the reports should be relevant to engineering and industry.

Pr. Yurii Loginov,

Vice Rector for Science and Research

Reshetnev University

IOP Conf. Series: Materials Science and Engineering **467** (2019) 011002 doi:10.1088/1757-899X/467/1/011002

Peer review statement

All papers published in this volume of *IOP Conference Series: Materials Science and Engineering* have been peer reviewed through processes administered by the proceedings Editors. Reviews were conducted by expert referees to the professional and scientific standards expected of a proceedings journal published by IOP Publishing.

IOP Conf. Series: Materials Science and Engineering **467** (2019) 012001 doi:10.1088/1757-899X/467/1/012001

Estimating the Effectiveness of Hydrodynamic Treatment of Wood for Producing Wood Boards without Adhesives

V N Ermolin[1], M A Bayandin[2], S N Kazitsin[3], A V Namyatov[4]

[1]Professor, Reshetnev Siberian State University of Science and Technology, 660037, Krasnoyarsk, Russia
[2]Associate Professor, Reshetnev Siberian State University of Science and Technology, 660037, Krasnoyarsk, Russia
[3]Assistant, Reshetnev Siberian State University of Science and Technology, 660037, Krasnoyarsk, Russia
[4]Aspirant, Reshetnev Siberian State University of Science and Technology, 660037, Krasnoyarsk, Russia

E-mail: mihailbayandin@yandex.ru

Abstract. This paper suggests the water retention value (WRV) as a criterion of estimation of wood pulp produced through treatment in the hydrodynamic disperser. Dependences between the WRV indicator and the properties of boards made of this pulp by the hot pressing method without using adhesives have been identified. As the WRV increases, bending and breaking strength of boards perpendicular to the sawn face grows, and the swelling value decreases.
Key words: wood boards, hydrodynamic treatment, sawdust, cavitation, strength, grinding, water retention value, swelling.

1. Introduction

Wood boards are widely used in the development of the human environment. Therefore, ecological indicators of boards become very important and topical under modern conditions. One of the main priorities is to improve these indicators while upgrading wood board manufacturing technologies. A possible solution to this problem is to develop board production methods without using adhesives.

The previous studies [1] indicate a possibility of formation of autoadhesion interactions between wood particles that have been preliminarily treated in the hydrodynamic disperser. The process of pressing of wood pulp produced by this method allows manufacturing boards without using adhesives. The density of boards may vary from 300 to 1,200 kg/m^3. Mechanical properties of these boards, at comparable density, are highly competitive with MDF, WCB (wood chipboards), WFB (wood fiber boards) and have higher water resistance. In addition, original raw materials for producing such boards are milled wood waste (sawdust, chips) the rational use of which is an important problem. This fact proves the potentials of the comprehensive study of this wood treatment method.

One of important questions is to choose a criterion of estimation of hydrodynamic wood treatment results on the basis of which it is possible to forecast the properties of boards.

The analysis of the conducted studies gives ground to suppose that the interaction between particles of hydrodynamically treated wood pulp takes place due to the physicochemical interaction between contacting surfaces. Thus, the more the surfaces are subject to contact, the stronger the autoadhesion interaction and the higher the strength of boards. As it is known, grinding increases an area of boundary surfaces. But according to the studies conducted with using wood dust, simple mechanical grinding brings no desired results. Also, particles should be fibrillated. This will allow us to increase the number of available boundary surfaces considerably. Due to a cavitation effect, hydrodynamic treatment provides the fibrillation of wood particles.

According to many researchers [5], the strength of paper depends mainly on the fibrillation of cellulosic fibers. In order to estimate a fibrillation degree in the paper industry, it is recommended to use the water retention value (WRV). This indicator is better than the Schopper-Riegler freeness value and characterizes the capability of fibers to form inter-fiber connections and produce tough paper [4]. According to this, we have accepted the WRV indicator in order to estimate the wood pulp produced by treatment in the hydrodynamic disperser.

IOP Conf. Series: Materials Science and Engineering **467** (2019) 012001 doi:10.1088/1757-899X/467/1/012001

2. Materials and methods

The special studies have been initially conducted in order to identify the dependence between the WRV indicator and the duration of treatment in the laboratory hydrodynamic disperser. Softwood sawdust has been used for conducting the experiments. The treatment has been carried out with hydromodulus 1:10. The water retention value has been determined according to the Jayme method [3] (Figure 1). Then, using the same technology, there has been carried out the hot pressing of treated wood pulp with the different WRV indicators. The pressing with the set density of 800 kg/m^3 has been carried out in the laboratory press (brand) at the temperature of 190^{0}C, the specific duration of 2.5 min/mm and the specific pressure of 3.5 MPa. Upon pressing, the boards have been kept in laboratory conditions for 72 hours and then have been subject to physical-mechanical tests. The results are given in Figure 2.

3. Results and discussions

The results of the study of the wood pulp treatment duration influence on the water retention value are given in Figure 1.

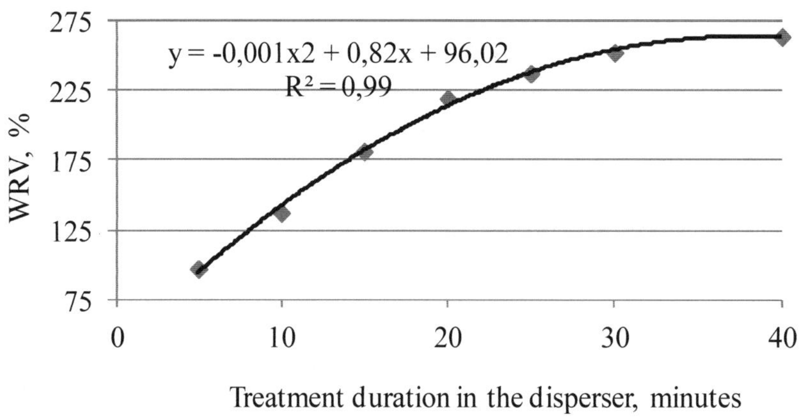

Figure 1. Dependence of the water retention value on the number of passes through the disperser.

Figure 1 shows that there is correlation dependence in the form of polynom between the treatment duration in the disperser and the WRV. As the treatment duration increases, the water retention value grows, in case of sawdust it is 97.1%. As a result of the treatment, it increases 2.8 times and reaches the value of 264%. The most intensive growth of the WRV is observed during the first 30 minutes, and then it does not change considerably, which indicates a decrease in the effectiveness of further treatment. It can be explained by the destruction of a fibrillar structure and the destruction of components of a cellular wall of wood. This corresponds to the opinion of the authors [4] who have studied mechanical grinding of cellulose.

The results presented in Figure 2 indicate that the properties of boards depend substantially on the WRV indicator of initial wood pulp. Ultimate bending strength (Figure 2a) and ultimate tensile strength perpendicular to the sawn face (Figure 2б), when the water retention value grows, increase by 2.15 and 2.5 times respectively. The swelling indicator has an inverse dependence and decreases from 33.2% to 6.9%. A sharp increase in indicators of boards is observed after increasing the WRV by more than 220%. For example, when the WRV varies from 220 to 270%, there is an increase in ultimate bending strength of boards from 13.05 to 25.68 MPa, breaking strength perpendicular to the sawn face from 0.54 to 1.02 MPa. And the swelling value of boards decreases from 21.7 to 6.9 %.

The analysis of the obtained results shows that hydrodynamic treatment allows making ecologically safe boards with high rates of properties of wood waste without using adhesives.

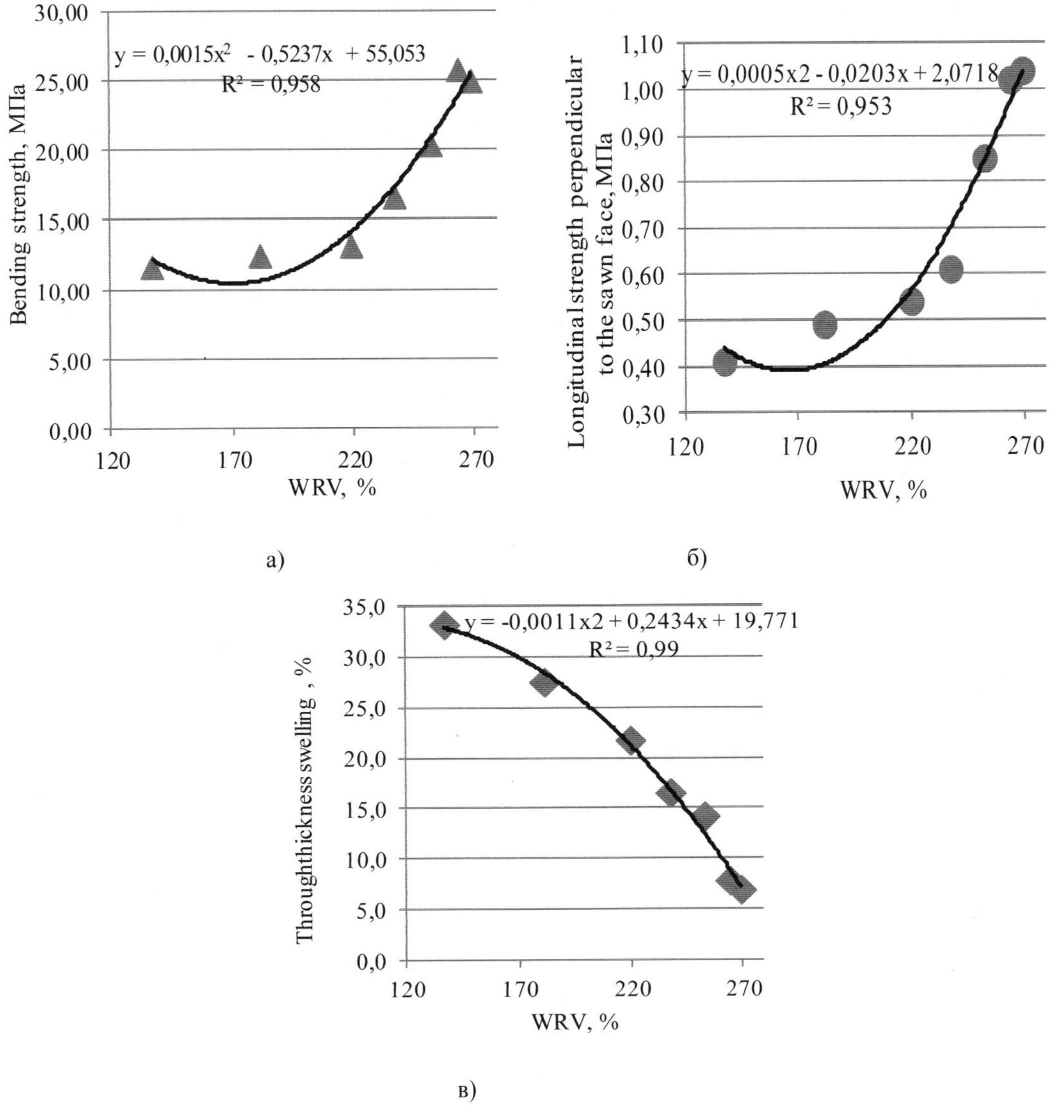

Figure 2. Dependences of physicomechanical properties of boards on the water retention value (WRV).

4. Conclusions

High rates of autoadhesion interaction between wood particles can be reached due to their fibrillation.

The WRV indicator may be used as a criterion of estimation of wood pulp produced by means of hydrodynamic treatment. There are correlation dependences between this indicator and the properties of produced boards.

The highest rates of properties are reached within the range of variation of the WRV from 220 to 270%.

The reported study was funded by Russian Foundation for Basic Research, Government of Krasnoyarsk Territory, Krasnoyarsk Region Science and Technology Support Fund to the research project №17-48-240553

References

[1] Bayandin M A, Ermolin V N, Kazicin S N 2015 Influence on the formation fine fraction properties of wood-based panels without connections. *Conifers of the boreal area 33* No **3-4**. pp 182-185

[2] Ermolin V N, Bayandin M A, Kazitsin S N Mechanical Activation of Wood for Adhesive-free board Production IOP Conf. Series: Materials Science and Engineering 155 (2016) 012038 doi:10.1088/1757-899X/155/1/012038
[3] Jayme G 1961 "Zellstoff und Papier" **11** pp 432-438
[4] Smolin A S, Bisalski M, Shabel C, Shabiev P O 2011 Influence of grinding and fractionation on electrosuperficial properties of cellulosic hydrosuspensions *Chemistry of herbal raw materials* No **3** pp 183-192
[5] Paulapuro H and Gullichsen J Papermaking science and technology 1998 – 2002, in 19 v., FAPET/TAPPI, Helsinki

IOP Conf. Series: Materials Science and Engineering **467** (2019) 012002 doi:10.1088/1757-899X/467/1/012002

Research of the paper-forming properties of a fibrous material with the use of a tacking with curvilinear knives

V I Shurkina[1], R A Marchenko[2], Yu D Alashkevich[3]

[1]Senior Teacher, Reshetnev Siberian State University of Science and Technology 660037, Krasnoyarsk, Russia

[2] Associate Professor, Reshetnev Siberian State University of Science and Technology 660037, Krasnoyarsk, Russia

[3] Chief Researcher, Institute of Chemistry and Chemical Engineering of Siberian Brunch of Russian Academy of Science, 660036, Krasnoyarsk, Russia

[1]Welta0007@mail.ru, [2]r.a.marchenko@mail.ru, [3]chem.@icct.ru

Abstract. This article considers the influence of a tacking with curvilinear knives on the quality of sulfated bleached softwood pulp crushing, certain paper-forming characteristics of the paper pulp are determined as well.

One of the efficient ways of increasing the quality of the finished products without modernizing a production line and as a consequence without raising production costs is crushing improvement due to the use of crucially new types of tacking [1].

At the Department of Industrial Engineering Machines and Apparatuses, Siberian State Aerospace University, a tacking with curvilinear knives was developed (Figure 1) [2]. This tacking improves the quality of wood pulp treatment, as well as performance and power consumption reduction.

It is important to study the influence of this tacking shape on the crushing quality of sulfated bleached softwood pulp, as well as to determine certain paper-forming characteristics of paper pulp, such as average fiber length, moisture retention capacity, interfiber bonding strength.

While analyzing the experimental data, the most efficient parameters of crushing for a tacking with curvilinear knives were chosen: paper pulp concentration of 3%, rotor speed of 2000 rpm and clearance between rotor and stator disks of 0.1 mm.

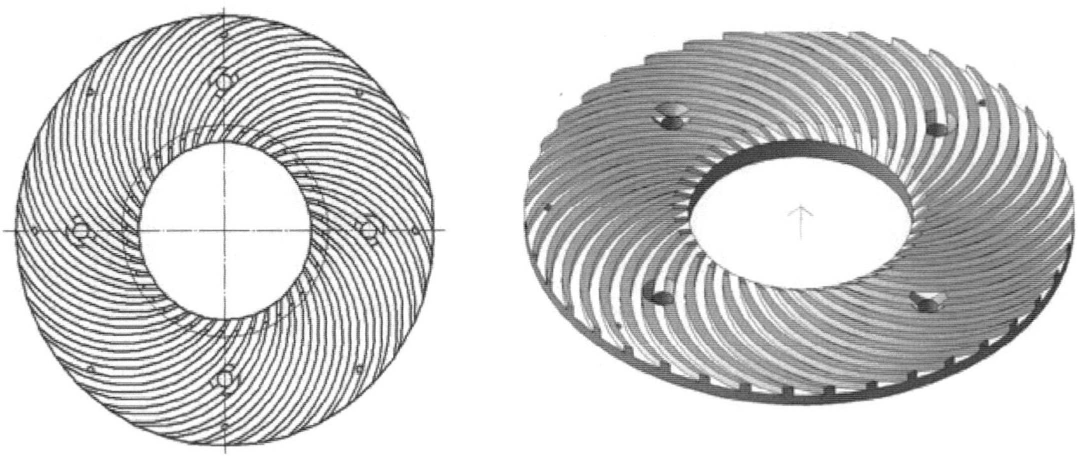

Figure 1. Tacking with curvilinear knives used for half-finished fiber products

For comparison with a curvilinear tacking, a standard eight-section knife tacking with an angle of knife crossing of 45° [3] and a force applying tacking without skew bevel were chosen [4].

Table 1 shows the main paper-forming characteristics of paper pulp when various designs of tacking are used.

IOP Conf. Series: Materials Science and Engineering **467** (2019) 012002 doi:10.1088/1757-899X/467/1/012002

Table 1. Main paper-forming characteristics of paper pulp and physical and mechanical characteristics of finished products when various design of tackings at 50 °SR are used.

Paper-forming characteristics of paper pulp	Tacking with curvilinear knives	Force applying tacking without skew bevel	Standard eight-section knife tacking with an angle of knife crossing of 45°
Crushing duration, min	21	28	35
Average fiber length, mm	1.58	1.38	1.23
Half-finished fiber product's moisture retention capacity, %	522	501	486
Interfiber bonding strength, kPa	0.157	0.124	0.142

For descriptive reasons the Microsoft Excel program was used to make diagrams that show dependence of changes in paper-forming characteristics when tackings of various design are used (Figure 2).

Table 1 and Figure 2 show that when a tacking with curvilinear knives is used, crushing duration is the shortest, while when a standard tacking is used, crushing duration is the longest.

Figure 2 also shows that there is a tendency of increasing a crushing rate, as well as crushing duration is increased; moreover, it should be noted that the intensity of the crushing rate is higher, when a tacking with curvilinear knives is used, in comparison with a force applying tacking or a standard knife tacking.

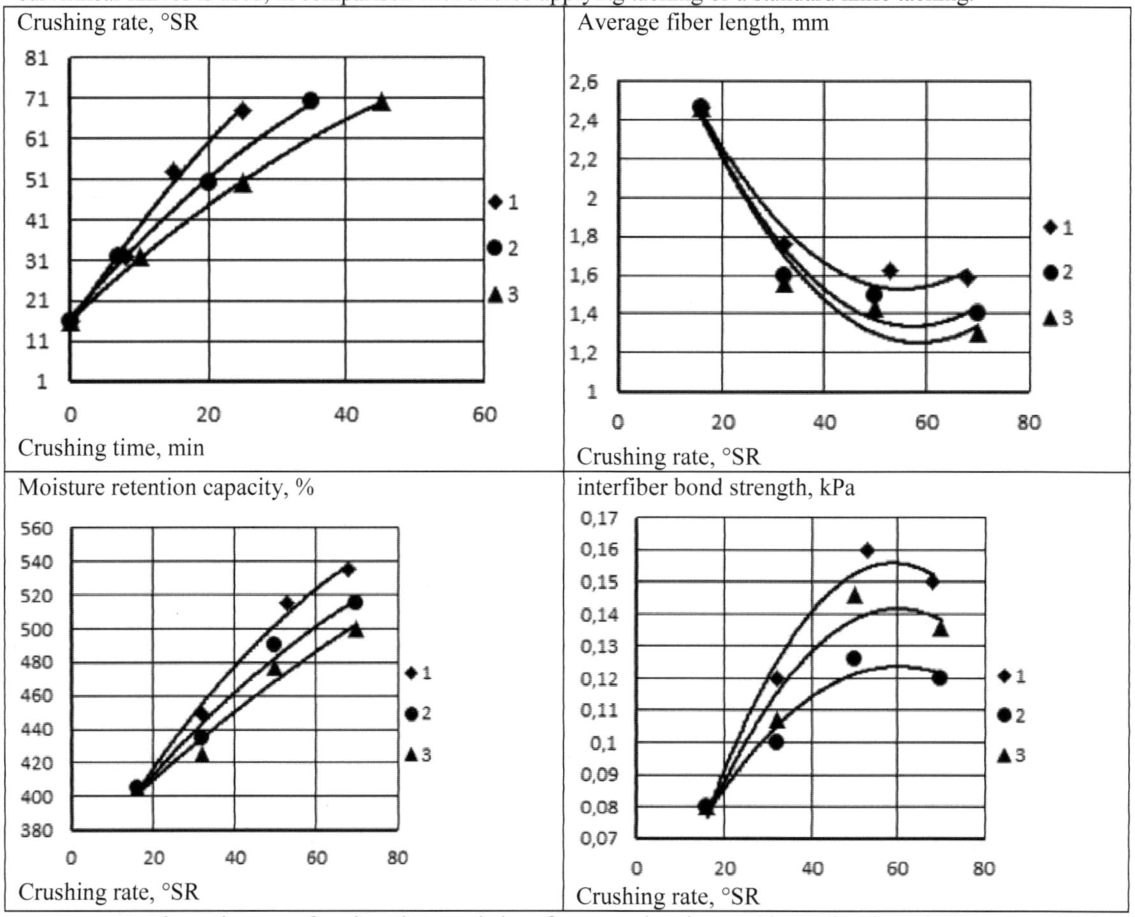

Figure 2. The main paper-forming characteristics of paper pulp when tackings of various designs are used.

The fact that crushing to a certain size by a curvilinear knife tacking demands less time positively affects power consumption. Besides, the studied tacking shows an increase in the most of physical and mechanical characteristics in comparison with the figures of a tacking of another shape.

Dependence (fig.2) of average fiber length on the crushing rate °SR and the type of tacking indicates the tendency of decrease in average fiber length while the crushing rate is increasing, which corresponds to the acknowledged data from literature sources [5].

The analysis of the diagrams shows that the curves irrespective of the design of a knife tacking are of equal qualitative character and are parabolic curves. However, quantitative values are different. When a tacking with curvilinear knives is used, fibers are shortened less intensively, which proves higher fibrillation of fibers, in comparison with tacking with rectilinear knives.

Table 1 and Figure 2 show that the highest rate of water retaining capacity belongs to paper pulp that is crushed by a tacking with curvilinear knives. As it is known, an increase in water-retaining capacity is characterized by fiber surface expansion, which in its turn contributes to a better contact and connection of separate fibers in a paper sheet [4].

The highest value of interfiber bonding strength in comparison with a rectilinear knives tacking is achieved when using a tacking with curvilinear knives.

Conclusions

Upon analyzing Table 1 and Figure 2, it may be concluded that a tacking with curvilinear knives is effective in terms of paper-forming characteristics of paper pulp. The obtained data confirm the efficiency of using a tacking with curvilinear knives in comparison with tackings of other designs.

References

[1] KarlssonMarkku, Malashenko Alex 2000 Developments in paper making technology with special emphasis on fine paper:*86th Annual Meeting of PAPTACPulp and Paper Technical Association of Canada*, Montreal, Febr. 1–3,.Prepr.IC – **I**. Montreal pp. 273–280.
[2] Alashkevich YuD, Kovalev V I, Kharin V F, Moukhachev A P 2007 Grinding tacking applicant and patent holder. Pat. 2307883. Russian Federation, p. 8.
[3] Nabieva A A2004 Assessment of influence and improvement of the main technological parameters of knife grinding machines: thesis.PhD in Technical Sciences. 05. 21. 03 p 182
[4] Kozhukhov V A2015 Grinding of half-finished fiber products in knife grinding machines by applying force thesis.PhD in Technical Sciences. 05. 21. 03 p 161
[5] Ivanov S N2006 *Paper technology*p 696

IOP Conf. Series: Materials Science and Engineering **467** (2019) 012003 doi:10.1088/1757-899X/467/1/012003

Application of the knife machine's operation elements with inertia bodies force to fiber

I A Voronin[1], Yu D Alashkevich[2], N S Reshetova[3]

[1]Associate Professor, Reshetnev Siberian State University of Science and Technology, 660037, Krasnoyarsk, Russia
[2]Chief Researcher, Chief Researcher, Institute of Chemistry and Chemical Engineering of Siberian Brunch of Russian Academy of Science, 660036, Krasnoyarsk, Russia
[3]Associate Professor, Reshetnev Siberian State University of Science and Technology, 660037, Krasnoyarsk, Russia

[1]ivan_voronin@list.ru, [2]chem.@icct.ru, [3]reshet@list.ru

Abstract. This article considers a shear force impact on fiber during the contact of inertia body's knives with drum knives.

In papers [1,2] it was discussed that the inertia force J was the main force providing fiber crushing during machine grinding with inertia motiving bodies. It was proved that the force of sliding friction and the force of rolling friction also influenced fiber crushing.

Hereby we consider conditions of intensive impact on the pulp of grinding machine operation elements. Under the force P_1 an inertia body located on the internal surface of the drum is tentatively deformed whereupon it leans not on one point, but on areas of bigger or smaller sizes [2]. It results in that during rolling of the body, the point A (figure 1) of the bearing reaction application moves a bit forward from the vertical line passing through the body's gravity center, while the action line of the bearing reaction R_A deviates a bit backward from this vertical line. At that the normal component RH = N of the bearing reaction compensates the inertia force J (i.e. RH = –J) and that not compensated tangential component R_{fr} of the bearing reaction is directed against the body movement and plays the role of rolling friction force T_r^{fr} or the force of friction against the drum wall [2].

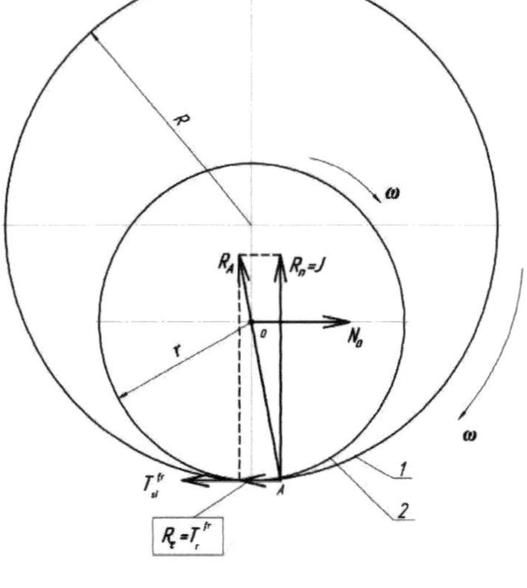

1 –grinding drum; 2 – inertia body; T_r^{fr} – force of rolling friction against the drum wall, H; R_n – normal component of bearing reaction, H; R_τ – tangential component of the bearing reaction, H; R_A – force of bearing reaction, H; T_{sl}^{fr} – force of sliding friction against the drum wall, H

Figure 1. Scheme of inertia bodies' impacts on fiber due to sliding friction and rolling friction forces

The modulus of rolling friction force against the drum wall T_r^{fr} is calculated using the formula:

$$T_r^{\text{fr}} = \frac{f_r \cdot P_1}{r}, \qquad (1)$$

where f_r is a coefficient of rolling friction equal to 0.0005 m [3];

r is a radius of an inertia body equal to 0.045 m.

The sliding friction force T_{sl}^{fr}, H, against the drum wall is calculated using the formula:

$$T_{sl}^{fr} = f_{sl} \cdot P_1 . \tag{2}$$

Hereby we consider the condition of the inertia body at various rotational speeds of the inertia body and the change in circumferential force N_{cir} [2]:

a) inertia body rolls without sliding, if $T_r^{fr} < N_{cir} < T_{sl}^{fr}$;

b) inertia body rolls with sliding, if $N_{cir} > T_{sl}^{fr}$

Hereby we determine the circumferential force affecting the inertia body N_{cir}, H

$$N_{cir} = \frac{P_{moto} \cdot \eta}{V_0}, \tag{3}$$

where P_{moto} is the power of motor, kW;

η is the efficiency coefficient of mechanical transmission form the engine [3];

V_0 is circumferential speed of rotation of the inertia body, m/sec.

Efficiency coefficient of mechanical transmission from the engine is calculated using the formula:

$$\eta = \eta_1^2 \cdot \eta_2 \cdot \eta_3 \cdot \eta_4, \tag{4}$$

where η_1 is roller bearing friction loss;

η_2 is friction losses in V-belt drives;

η_3 is friction losses in bevel drive;

η_4 is losses in shaft bearings of the apparatus drive.

$$V_0 = \frac{\pi \cdot D_c \cdot n}{60}, \tag{5}$$

where D_c is a diameter of the inertia body, m.

The results of calculation of forces affecting fiber are summarized in Table 1.

Table 1 shows that within the change in the rotational speed of the inertia body up to 883.5 rpm, the latter rolls down the inner drum surface generatrix with sliding as $N_{cir} > T_{CK}^{\text{тр}}$. With further increase in inertia body rotational speed, it rolls without sliding [3], since from that moment an increase in the value of the rotational speed of the inertia body causes a decrease in the value of circumferential force N_{cir} and an increase in the force of sliding friction against the drum wall T_{sl}^{fr}, which provides fulfillment of the condition $T_r^{fr} < N_{cir} < T_{sl}^{fr}$. At the same time, the inertia force J keeps incrementing together with the increase in the rotational speed of the machine's operation bodies.

IOP Conf. Series: Materials Science and Engineering **467** (2019) 012003 doi:10.1088/1757-899X/467/1/012003

Table 1. Forces affecting fiber in case rotational speed of the inertia body changes

The rotational speed of inertia body n, rpm	Inertia force J, H	Circumferential stress of inertia body N_{cir}, H	The force of rolling friction against the drum wall T_r^{fr} H	The force of sliding friction against the drum wall T_{cr}^{fr}, H
85.5	11.61	2497.05	0.13	2.32
114	20.64	1872.78	0.23	4.13
142.5	32.26	1498.23	0.36	6.45
171	46.45	1248.52	0.52	9.29
199.5	63.22	1070.16	0.71	12.64
228	**82.58**	**936.39**	**0.93**	**16.52**
256.5	104.51	832.35	1.17	20.90
285	129.03	749.11	1.45	25.81
313.5	156.12	681.01	1.75	31.22
342	185.80	624.26	2.09	37.16
370.5	218.05	576.24	2.45	43.61
399	252.89	535.08	2.84	50.58
798	1011.56	267.54	11.37	202.31
826.5	1085.10	258.32	12.19	217.02
855	1161.23	249.70	13.05	232.25
883.5	**1239.93**	**241.65**	**13.93**	**247.99**
912	1321.22	234.10	14.85	264.24

Taking into consideration that the crashing capability of the machine provides the maximum value of rotational speed of inertia bodies corresponding to 228 rpm, the latter move in the machine with sliding, which corresponds to the fulfillment of the condition $N_{cir} > T_{sl}^{fr}$.

Thus, if the machine functions within its technical characteristics with the maximum speed of inertia bodies rotation of 228 rpm, the dominant forces of impact on the fiber while grinding are the forces of inertia J and sliding friction T_{sl}^{fr} .

Hereby we consider the dependence of sliding friction and rolling friction forces on circumferential force (figure 2).

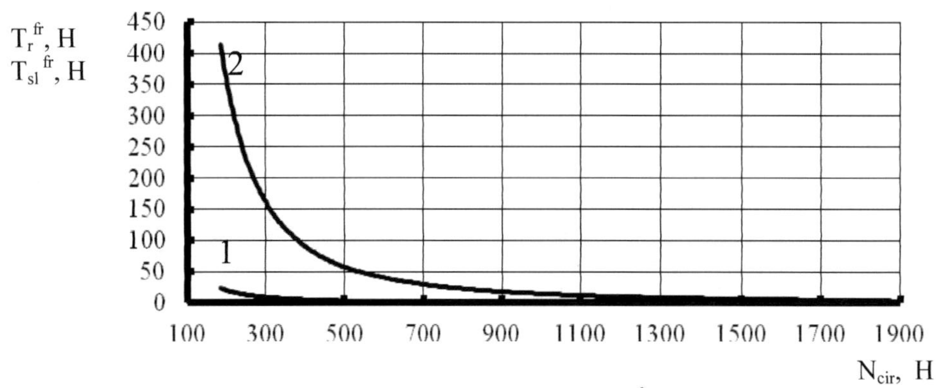

1 is the force of rolling friction against the drum wall T_r^{fr} H

2 is the force of sliding friction against the drum wall T_{sl}^{TP} , H

Figure 2. Dependence of sliding friction and rolling friction forces on circumferential force

Figure 2 presents the diagram which shows that when circumferential force increases, the force of rolling friction and sliding friction decrease. The force of sliding friction against the drum wall is the maximum force providing the maximum crushing effect on fiber. The diagram also shows that the decrease in circumferential force (increase in the rotational speed of inertia bodies) results in a sharp increment of sliding friction force. It may be supposed that when the rotation speed of the inertia body increments, the critical grinding process is slowing down.

Conclusions

When the inertia body moves in the grinding drum, fiber is mostly affected by the force of inertia J and sliding friction T_{sl}^{fr} against the drum wall, which is the maximum force providing the maximum crushing effect on fiber. The force of rolling friction T_r^{fr} exerts an insignificant effect.

References

[1] Alashkevich Yu D, Voronin I A, Kovalev V I, Reshetova N S 2009 Stock beating by a non-conventional method: *Vegetation stock chemistry* **2** pp 165-168

[2] Voronin I A 2011 Unconventional method of the fibrous suspension processing with the use of inertia bodies in the pulp and paper industry. synopsis of the thesis. PhD (Technical Sciences) p 22

[3] Anuriev V I 2001 Handbook of a designer and mechanical engineer p 920

[4] Flyate D M 1986 Paper technology p 680

IOP Conf. Series: Materials Science and Engineering **467** (2019) 012004 doi:10.1088/1757-899X/467/1/012004

Static and modal analysis of the construction of the precision large-sized antenna reflector from polymer composite materials

A Y Vlasov[1], N A Amelchenko[2], K A Pasechnik[3], M A Titov[4] and M S Serzhantova[5]

[1]Associate professor; Reshetnev Siberian State University of Science and Technology, 660037, Krasnoyarsk, Russia.
[2]Associate professor of mechanical engineering department, Reshetnev Siberian State University of Science and Technology, 660037, Krasnoyarsk, Russia.
[3]Engineer; Reshetnev Siberian State University of Science and Technology, 660037, Krasnoyarsk, Russia.
[4]Engineer; Reshetnev Siberian State University of Science and Technology, 660037, Krasnoyarsk, Russia.
[5]Associate professor; Reshetnev Siberian State University of Science and Technology, 660037, Krasnoyarsk, Russia.

E-mail: vlasov.anton@gmail.com

Abstract. The article presents a design of a precision large-size antenna reflector that made of polymer composite materials (PCM) based on carbon fibers. Such reflectors are used for operation in high frequency ranges, since they have a low coefficient of linear thermal expansion and a high modulus of elasticity. Therefore, the main task of this work is to design the geometric accuracy of the reflector working surface from composite materials with a diameter of more than 10 meters and a frequency range of 42.5-45.5 GHz.

The developed model of the reflector includes a power frame, segments of the reflecting surface and a hub. The power frame of the reflector consists of flat trusses supplemented with rods, so that during assembly a spatial construction with axial symmetry is formed. Segments are three-layer casings of polymer composite materials with filler.

The proposed model of the reflector was analyzed using the finite element method with boundary conditions: a wind load of 20 m/s in the opening of the reflector; impact of gravity on the reflector, oriented to the zenith. The wind load was modeled as a uniformly distributed pressure applied to the segments. The obtained mean square deviation (SDE) of the geometry and natural oscillation frequency of a reflector made from polymer composite materials based on carbon fibers is sufficient for operation of the satellite earth station in high radio-frequency ranges Ka, Q and V.

1. Introduction. Currently the systems which are located on HTS (High Throughput Satellite) satellites characterized by using the high frequency ranges, including *Ka* (18,2 – 21,2 GHz), *Q* (42,5 – 45,5 GHz) and *V* (50 – 70 GHz) [1]. Providing the broadband communication channels requires to make the precision large-sized antenna reflectors with a diameter being more 10 meters, and working in the radio frequency bands as *Ka*, *Q* and *V*. The task of designing is to achieve geometric accuracy of the reflector's working surface that expressed in terms of SD [2] from basic profile

$$\sqrt{1/n\sum_{i=1}^{n}\left(x_i-\overline{x}\right)^2},$$

where – n is a sample volume of reflector surface points, $\overline{x}$ – sample arithmetic mean, x_i – sample unit, as well as ensuring resistance to mechanical stress and climatic factors.

The precision large-sized antenna reflector examples are IRAM (30 meters и 15 meters telescope located in Spain and France), Tian Ma (65 meters, China, USA), CSO (10 meters, Hawaii), APEX (12 meters, Chili), ALMA (12 meters, Chili) [3-7]. Above mentioned projects examples are radio telescopes, that reflect SD surface vary from 0,015 to 0,03 mm at diameters to 12 meters, to 0,3 mm at diameters to 65 meters. But using such engineering solutions for satellite earth station reflector constructions is unacceptable because of high cost and complexity of maintenance.

IOP Conf. Series: Materials Science and Engineering **467** (2019) 012004 doi:10.1088/1757-899X/467/1/012004

2. Setting goal. Polymer composites based on carbon fibers have got low linear thermal expansion coefficient (LTEC) a high module of elasticity [8]. Object of this research is to develop large-size integrated dimensional reflector design that is made of polymer composites working in the range of 42,5 – 45,5 GHz with low value of the standard deviation of the reflecting surface and the high rigidity of the structure.

SD interval of allowed values for *Q* frequency range with wave length λ = 6,58 … 7,05 mm defines of interrelation $\sigma = \lambda/50$ that equals 0,13 mm in the case of zero-wind force and interrelation $\sigma = \lambda/12$ [9] that equals 0,55 mm in the case of wind force.

3. Model description. Developed reflector model includes three parts: load-bearing unit, reflecting surface segments, and hub [10, 11].

Load-bearing unit is a spatial truss configuration that is designed to provide rigidity, strength and temperature dimensional stability of reflective surface of antenna reflector (Fig. 1).

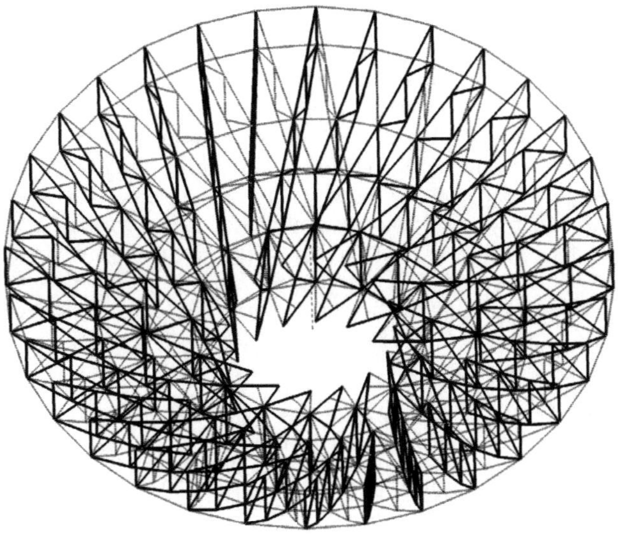

Figure 1. General view of the reflector truss

The developed reflector unit consists of flat trusses that simplify its production, installation and transportation. The flat trusses are constructed separately that provides high accuracy of assembly. The flat trusses together with additional connecting rods are installed at an angle 58°30′ relatively to the normal restored to the base of the reflector. The proposed solution to the task allows to produce the load-bearing unit consisting of a set of flat trusses of two types: five-row and three-row ones (Fig. 1, 2); they are completed by the rods, with their nodes connecting the flat trusses among themselves in the way that a spatial unit with axial symmetry is formed during the assembly.

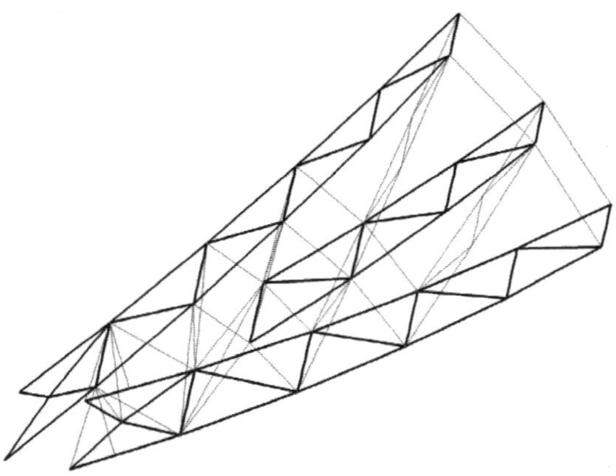

Figure 2. Regular element of the reflector truss
The thickened lines show flat radial trusses

IOP Conf. Series: Materials Science and Engineering **467** (2019) 012004 doi:10.1088/1757-899X/467/1/012004

The rods are made of carbon-fiber pipes with a diameter of 60 mm; there are two types: with a wall thickness of 1.4 mm and 2.8 mm. Strengthened pipes are located in the most loaded parts of load-bearing unit – in the first and second rows. While calculating the rod material is assumed to be isotropic with a modulus of elasticity of 70 hPa.

Shape and overall dimensions of reflecting surface segments determine the location of the nodal points of the load-bearing unit; this results in solving the optimization task to determine the spatial structure of the unit, which meets the operational requirements.

Segments are three-layered coats, front and back surfaces made of polymer composite materials, constructive styrofoam is used as a filler [12].

A hub is a welded steel unit of cylindrical shape that has interface attachment points to the support-rotary device.

4. Results. The proposed reflector model is analyzed by using the finite element method [13, 14] with following boundary conditions: wind load 20 m/s in reflector aperture, impact of gravity on zenith-oriented reflector. The load-bearing unit is designed by beam finite elements; segments are by superficial finite elements. The segments are rigidly connected with load-bearing configuration nodes. The fixing points of the structure to the hub are fixed by rigid sealing.

Wind load is modeled as evenly distributed pressure applied to the segments. Gravity is taken into account proportionally to the mass of the structural elements for each node of the finite element mesh.

SD calculation value without wind load is 0,16 mm including temperature-induced variations of 0,11 mm and gravitational ones of 0,05 mm. Under wind load the contribution to deformation is 0,28 mm and maximum SD calculation value increases to 0,44 mm. SD obtained values are in allowed deviation margin for Q frequency band. Figure 3 is a graph of the dependence of SD on the diameter of a reflector which is made of various structural materials, that results in the use of carbon fiber composites to obtain a large-sized dimensionally stable configuration.

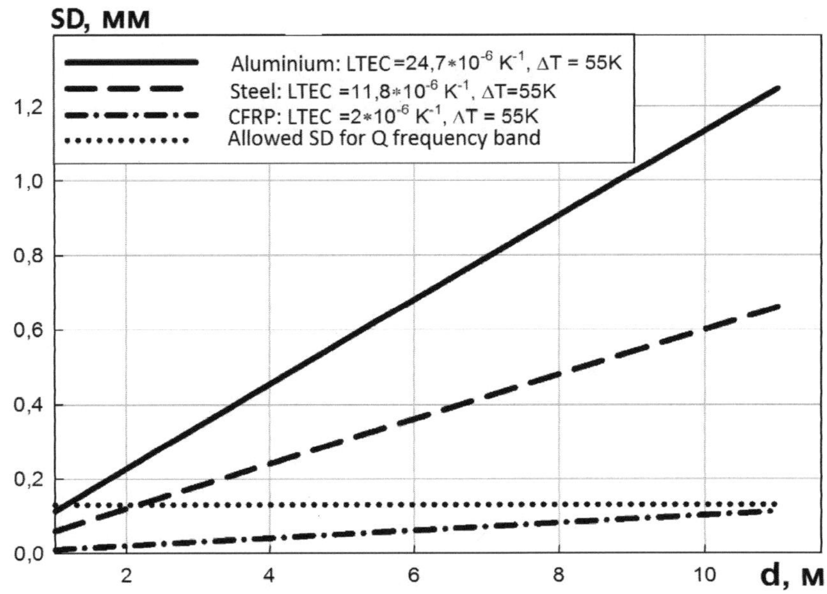

Figure 3. Standard deviation function graph

The modal analysis of the load-bearing unit demonstrates that the first self-resonant frequency is 47 Hz with a mass of the configuration of about 800 kg, while for analogous reflectors KAT-7 with a diameter of 12 m, the natural frequency is 4 Hz [15]. This indicates the high rigidity of the designed configuration.

5. Conclusion. Developed model of large-sized antenna reflector of polymer composite materials based on carbon fibers has a high rigidity of configuration and reflective surface SD in interval from 0,13 mm to 0,55 mm that provides a stable operation of a satellite earth station in the range of 42,5 – 45,5 HGz.

Acknowledgment. The work was carried out during the implementation of a comprehensive project with the financial support of the Government of the Russian Federation (Ministry of Education and Science of Russian Federation). Contract No. 02. G25.31.0147.

References

[1] Testoedov N A and Kuzovnikov A V 2017 Prospects and priorities for the development of information satellite systems. *Issledovaniya naukograda.* **1 19** 7–10
[2] Beklemishev D V 2008 *The Course of Analytic Geometry and Linear Algebra (Moscow: FIZMATLIT)*
[3] Vereshchagin E M 1964 *Antennas and propagation of radio waves (Moscow: Voenizdat)*
[4] Vud P 1984 *Analysis and design of mirror antennas (Moscow: Radio i svyaz)*
[5] Scott C 1990 *Modern Methods of Reflector Antenna Analysis and Design (London: Artech House Boston)*
[6] Cheng J 2009 *The principles of astronomical telescope design (New York: Springer)*
[7] Khristiansen U and Khegbom I 1988 *Radiotelescopes (Moscow: Mir)*
[8] Daskovskiy M I, Doriomedov M S and Skripachev S Y 2016 Systematization of basic factors obstacling the implementation of polymer composite materials in Russia (Review) *Trudy VIAM* **5 41** 44–52
[9] Frolov O P and Vald V P 2008 *Mirror antennas for satellite earth stations* (*Moscow: Goryachaya Liniya – Telekom)*
[10] Vlasov A Y, Pasechnik K A and Martynov V A 2015 Design – Technological Aspects of Creating Precision Products of Complex Form from Polymer Composite Materials. *Vestnik SibGAU* **2 17** 460–466
[11] Vlasov A Y, Titov M A and Pasechnik K A 2015 Development of the optimal constructive-power scheme of a six-segment offset antenna reflector using polymer composite materials. *Mezhdunarodnyi nauchno-tekhnicheskii zhurnal. Uspekhi sovremennoi radioelektroniki* **10** 28–33
[12] Vlasov A Y, Pasechnik K A and Martynov V A 2015 Development of the technology of partial curing of thin-walled casings when creating a structure of polymer composite materials *Vestnik SibGAU* **4 16** 918–924
[13] Vlasov A Y, Lopatin A V et al 2013 Development of precision antenna reflectors from polymer composite materials: finite element simulation of the structure *Vestnik SibGAU* **3 49** 73–78
[14] Vlasov AY , Pasechnik K A et al 2014 Description of the ranges of changes in key parameters that ensure the stability of the technological process, the information of complex shapes from PCM by the RTM method *Vestnik SibGAU* **4 56** 197–201
[15] Foley A R Engineering and Science Highlights of the KAT-7 Radio Telescope *Mon. Not. R. Astron. Soc.* **000** 1–17

IOP Conf. Series: Materials Science and Engineering **467** (2019) 012005 doi:10.1088/1757-899X/467/1/012005

Effect of oxygen dissolved in germanium on defect formation and optical properties of single crystals

A F Shimanskii [1,3], T O Pavluk[2], S A Kopytkova[2], A N Gorodishcheva[1]

[1]Reshetnev Siberian State University of Science and Technology, 660037, Krasnoyarsk, Russia
[2]JSC "Germanii", 660027, Krasnoyarsk, Russia
[3]Siberian Federal University, 660041 Krasnoyarsk, Russia

E-mail: shimanaf@mail.ru

Abstract: Oxygen impurities and optical properties of Sb-doped germanium single crystals were investigated with Fourier transformed infrared spectrometry. The effect of oxygen on crystal structure was shown. The effects of annealing in the temperature range 623–723 K on the behavior of oxygen dissolved in germanium and on optical properties of single crystals have been investigated. It was found that annealing at lower oxygen partial pressure leads to the formation of thermal donors (TD) on the basis of dissolved oxygen atoms. It was established that TD formation and the decrease of unbound oxygen concentration is accompanied by the increase in temperature stability of the optical properties of crystals.

Germanium single crystals have extensive application in nanoelectronics, detector engineering and IR optics. In photovoltaic germanium is used as a substrate material for solar cells GaInP/GaInAs/Ge type with the conversion efficiency up to ~ 39 % [1–5]. It requires low impurity dislocation-free Ge-crystals since dislocations and uncontrolled impurities can cause a mismatch between the lattice parameters of germanium and $A^{III}B^{V}$ compounds, impeding the growth of high-quality epitaxial layers on germanium [1, 3–4].

High carrier mobility in dislocation-free Ge promotes its application in the radiation-hardened power MOSFET transistors and other fast digital electronics of the space class [6].

Dislocations and foreign impurities limit the use of germanium in infrared optics, since they change the optical properties of germanium [5].

Recent studies have proved the oxygen to be one of the main impurities that affects structural perfection, electrical and optical properties of germanium single crystals and operational characteristics of Ge-based electronics [7–17].

Oxygen atoms dissolved in germanium are located at interstitial positions O_i. This defect in the simplest model may be regarded as a nonlinear symmetric quasi-molecule Ge–O–Ge, with vibration modes ν_1, ν_2 and ν_3. It is the general practice to determine oxygen concentration in the crystals by the absorption peak at 856 cm^{-1} in the infrared spectra, which is identified with asymmetrical vibration mode ν_3 [7–9]. In the recent research [10, 11] the oxygen peak at 856 cm^{-1} has been reported to have a "shoulder" at ν that equals 843 cm^{-1}, which is also identified with O_i vibrations.

It should be noted, that oxygen concentration in crystals being investigated was ~10^{17} cm^{-3} and more [7–11].

Our paper [14], as well as other papers [12, 13] show, that there is no band at 856 cm^{-1} in infrared spectrum with low oxygen (about 10^{16} cm^{-3} and less). In the interval from 800 to 900 cm^{-1} the band at 843 (841) cm^{-1} is observed. This band is interpreted as an "oxygen band" by us and by other authors [12, 13].

Oxygen is known to exist in atomic and bond state, e.g. in the form of precipitates GeO_x which are formed either during crystal growth or due to decomposition of supersaturated solid solution of oxygen in the process of post-growth annealing and cooling. The oxygen content in precipitates can achieve 20 % of its total concentration. It is established that the annealing of germanium at the temperatures from 623 to 723 K leads to the formation of the precipitates with electric activity, named thermal donors (TD). The majority of modern TD models are based on the idea that these centers are complexes with an electrically active core and different bounded numbers of oxygen atoms [4, 15–18].

Oxygen concentration in the crystals being investigated in papers [4, 15–18], as well as in papers [7–11] was ~10^{17} cm^{-3} and more. Thus, Ge was sufficiently oxygenated.

In this regard, the aim of our work is to investigate the behavior of oxygen, dissolved in Ge and its effect on the structure and optical properties of single crystals with lower oxygen content (about 10^{16} cm^{-3}).

The studies were made on Sb-doped Ge single crystals with a donor concentration between $1.1{\bullet}10^{15}$ and $7.0{\bullet}10^{14}$ cm^{-3}, corresponding to a specific electric resistivity in the range of 2 to 4 Ωm•cm. The material with dislocation content about ~ 10^4 cm^{-2} has absorption coefficient α of 0.15–0.20 cm^{-1} at the wavelength of 10.6 μm [4, 5].

Ge crystals were grown by the Czochralski method. The annealing of crystals at $P_{O_2} \approx 10^{-3}$ Pa and temperature interval from 623 to 723 K was conducted in Krypton (6N).

Polished plane-parallel plates with the thickness of 1 cm were prepared to study optical properties with Fourier transformed infrared spectrometry. The infrared measurements were performed in the 700–1500 cm^{-1} spectral range with SPECTRUM BXII spectrometer. The optical density measurement error was no more than ± 0.001.

The oxygen concentration $[O_i]$ was calculated from the measured amplitude of the absorption band at 843 cm^{-1} using the formula

$$[O_i] = 1.05 \cdot 10^{17} \left(\frac{2.3D}{d} \right), \tag{1}$$

where D – is the optical density relative to the baseline; d – is the sample thickness; $1.05 \cdot 10^{17}$ cm^{-2} – is the calibration factor [4].

The influence of isothermal annealing on the optical properties of single crystals was determined by the optical density value and absorption coefficient α at a wavelength of 10.6 μm. For IR measurement with the temperature rising up to 60°C the heat device ensuring a stable sample thermostating with accuracy ± 0.1 K was used.

The table 1 lists the changes in standard Gibbs energy for reactions involving germanium and its oxides and the corresponding pressure logarithm of germanium dioxide dissociation $\lg P_{O_2(GeO_2)}$

Table 1 Changes in the standard Gibbs free energy of the reactions involving germanium and oxygen and the corresponding germanium dioxide pressures logarithm

T [K]	923	1073	1223	1433
		$2GeO_{(r)} = GeO_2 + Ge$		
ΔG_T^o [J/mol]	−165250	−112930	−61660	0
		$GeO_2 = Ge + O_2$		
ΔG_T^o [J/mol]	+401350	+375540	+345630	+303540
$\lg P_{O_2(GeO_2)}$	−22.75	−18.21	−14.78	−11.10

The results of calculations were used to determine the regions of existence of germanium and GeO_2 in the temperature range of 923–1433 K (Fig. 1).

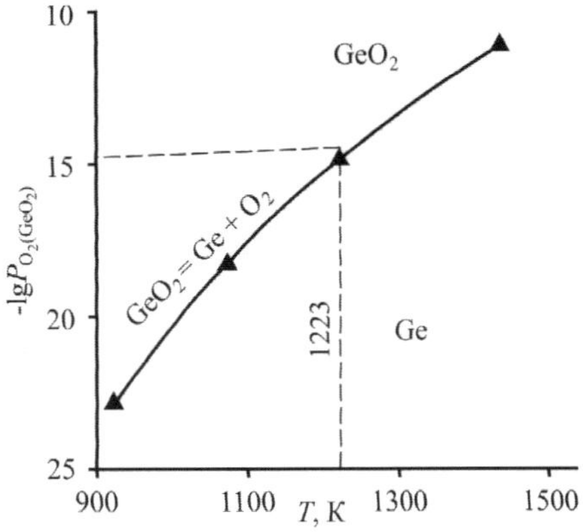

Figure 1. Germanium dioxide dissociation pressure logarithm and the Ge and GeO_2 existence regions

In the process of a single crystal growth germanium melt was heated to some degrees above the melting point (T_{melt} = 1210 K), the melt temperature was ~1223 K. Fig. 1 shows that germanium dioxide GeO_2 can be in balance with liquid germanium at this temperature (its dissociation pressure $P_{O_2(GeO_2)}$ at T =1223 K is 1.66•10^{-15} atm (1.66•10^{-10} Pa)). This means that germanium is oxidized when partial pressure of gaseous oxygen is higher than this value, making GeO_2 emerge on the surface of the melt. GeO_2 microparticles in the melt can be captured by the growing crystal to form heterogeneous inclusions, thus disturbing the homogeneity of the germanium crystal structure and causing dislocation formation. Dislocation formation, affected by oxygen in molten Ge was also mentioned in other papers [7].

If the oxygen partial pressure is lower than 1.66•10^{-15} atm (1.66•10^{-10} Pa), O_2 is completely soluble in germanium. In the process of crystal growth a solid solution is formed from melt, where oxygen is present as interstitial atomic O_i, which can also affect the structure and properties of Ge crystals. In view of this, the investigation of dissolved oxygen and its effect on optical properties of single crystals was conducted.

Fig. 2 shows IR absorption spectra of Sb-doped Ge single crystals with specific electrical resistance of 3 Ωm•cm at room temperature and heated up to higher temperatures in the range of wave numbers from 700 to 1500 cm^{-1}.

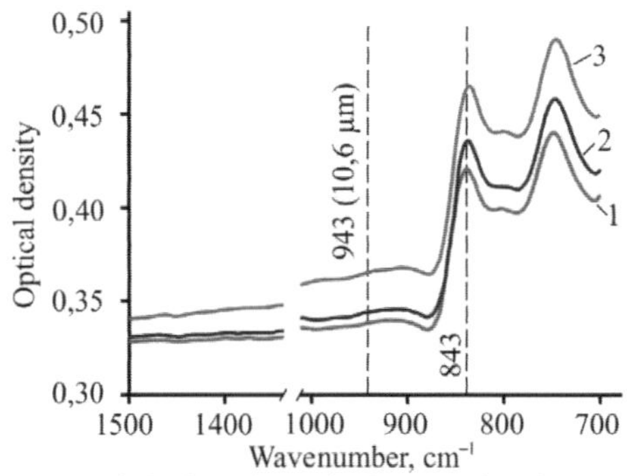

Figure 2. IR spectra of Sb-doped Ge single crystals with specific resistance of 3 Ωm•cm (1 – 293 K; 2 – 313 K; 3 – 333 K)

Judging by the value of optical density in the maximum of a band at 843 cm^{-1} regarding baseline the concentration of atomic oxygen in the investigated Ge crystal was determined. It was ~ 1.10•10^{16} cm^{-3}.

The effect of annealing in the temperature range from 623 to 723 K with partial pressure of oxygen P_{O_2} ≈ 10^{-3} Pa during the period of 24-90 hours on the behavior of interstitial oxygen and optical absorption of Ge crystals was investigated.

The experiment has proved that annealing leads to the reduction of band intensity 843 cm^{-1} that corresponds to the content decrease of interstitial atomic O_i as a component of quasi-molecules Ge–O–Ge. The 90-hour period annealing at $P_{O_2} \approx 10^{-3}$ Pa and temperature 673 K in Krypton (6N) leads to the oxygen band 843 cm^{-1} intensity reduction on 5 % (Fig. 3).

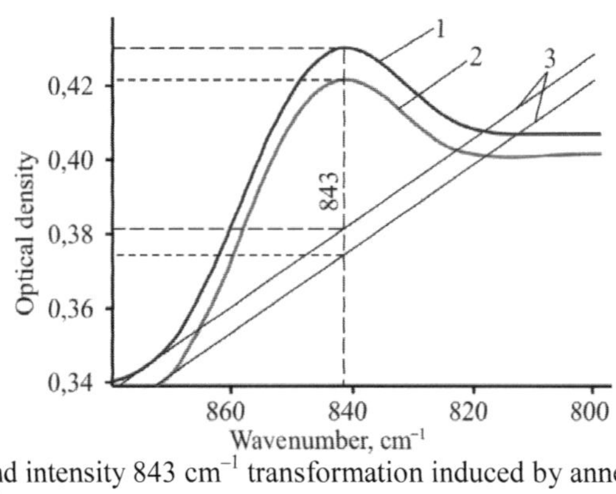

Figure 3. Band intensity 843 cm^{-1} transformation induced by annealing at 673 K and $P_{O_2} \approx 10^{-3}$ Pa (90 hrs) (1 – before annealing; 2 – after annealing; 3 – baselines)

The phenomenon observed in a similar way in papers [11, 18] can be explained by the formation of thermal donors (TD) based on dissolved oxygen in the germanium crystal lattice during annealing. Thus, the part of dispersed oxygen is bound in TD structure due to annealing.

It was found that the germanium annealing at low oxygen partial pressure ~ 10^{-3} Pa results in the change of optical characteristics of crystals, such as optical density and absorption coefficient.

The data in Fig. 2 show that the optical density of the sample with electrical resistance of 3 Ωm•cm at the room temperature and wavelength 10.6 μm is 0.337. When the temperature increases up to 333 K the optical density increases to 0.365 accordingly. That *D* change matches the absorption coefficient increase from 0.017 to 0.065 cm^{-1}. This temperature destabilization of germanium optical properties impedes its application in IR optics at overheating to 318 K.

Table 2 shows the optical properties of the samples annealed at temperature 673 K and $P_{O_2} \approx 10^{-3}$ Pa during 90 hour-period.

Table 2 Effect of annealing on optical properties of Ge single crystals

№ п/п	Electrical resistance [Ωm•cm]	Before annealing				After annealing at 673 K and $P_{O_2} \approx 10^{-3}$ Pa during 90 hrs	
		Optical Density	Absorption coefficient [cm^{-1}]	Optical Density	Absorption coefficient [cm^{-1}]	Optical Density	Absorption coefficient [cm^{-1}]
		Temperature 293 K		Temperature 333 K		Temperature 333 K	
1	2	0.339	0.020	0.364	0.062	0.359	0.054
2	2.5	0.339	0.020	0.366	0.066	0.364	0.062
3	3	0.337	0.017	0.365	0.065	0.360	0.056
4	3.5	0.336	0.015	0.370	0.073	0.361	0.057
5	4	0.334	0.010	0.366	0.067	0.362	0.059

It was found that optical density of these samples at 333 K is decreased from 0.365 to 0.360 after annealing at low oxygen partial pressure and the absorption coefficient is decreased from 0.066 to 0.056cm^{-1} accordingly. On average the absorption coefficient of Ge single crystals at 333 K decreases by 13% for the samples with electrical resistance in the range from 2 to 4 Ωm•cm as the result of annealing. The annealing influence on the optical properties of crystals at the room temperature was insignificant.

The experimental data show that the annealing of Ge single crystals at 673 K and $P_{O_2} \approx 10^{-3}$ Pa provides the temperature stability of their optical properties. It is assumed that the temperature stability improvement of the optical properties is due to oxygen concentration decrease and TD formation.

Oxygen in Ge is one of the main impurities, which defines the structural perfection and properties of crystals.

When the partial pressure of oxygen in the process of growing is higher than $1.66 \cdot 10^{-15}$ atm ($1.66 \cdot 10^{-10}$ Pa) germanium is oxidized with oxide GeO_2 on the surface. Microparticles GeO_2 can be captured by a growing crystal from melt and it leads to dehomogenization of crystal structure and formation of dislocations.

Dissolved oxygen in its turn affects optical properties of crystals.

Annealing of Ge crystals at $P_{O_2} \approx 10^{-3}$ Pa and temperature 673 K leads to the oxygen band 843 cm^{-1} intensity reduction due to TD formation. The decrease of unbound oxygen and TD formation are accompanied by the improvement of the temperature stability of the crystal optical properties.

The study was financially supported by the Russian Foundation for Basic Research, Government of Krasnoyarskii krai, and Krasnoyarskii krai foundation for the support of scientific and scientific-technological activities under project no. 16-43-240719.

References

[1] Dimroth F, Kurtz S 2007 *MRS bulletin* **32 (03)** p 230–235
[2] Rakwal D, Bamberg E 2009 *Journal Mater Process Technol* **209 (8)** p 3740-3751
[3] Hegedus S S, Luque A 2003 Handbook of photovoltaic science and engineering (John Wiley & Sons)

[4] Claeys L, Simoen E 2007 Germanium-based technologies: from materials to devices (Elsevier, Oxford)
[5] Depuydt B, Theuwis A, Romandic I 2006 *Mater Sci Semicond Process* **9 (4)** p 437–443
[6] Chroneos A, Vovk R V 2015 *Journal of Materials Science: Materials in Electronics* **26 (10)** p 7378–7380
[7] Taishi T, Ise H, Murao Y, Osawa T et al. 2010 *Journal Cryst Growth* **312 (19)** p 2783–2787
[8] Pajot B, Clauws P 1987 *The Proc of the 18th Int Conf on the Phys Of Semicond* **2** p 911–914
[9] Clauws P 1996 *Mater Sci Eng B Solid State Mater Adv Technol.* **36 (1)** p 213–220
[10] Pajot B, Clerjaud B 2013 Optical absorption of impurities and defects in semiconducting crystals electronic absorption of deep centres and vibrational spectra (Springer, Berlin)
[11] Inoue K, Taishi T, Tokumoto Y, Murao Y, Kutsukake K 2013 *Appl. Phys.* **113 (7)** p 1–5
[12] Kaplunov I A, Rogalin V E, Gavalyan M Yu 2015 *Optic and spectroscopy* **118 (2)** p 240–246
[13] Seref K, Romandicb I, Theuwisb A 2006 *Mater.s Sci. in Semicond. Proc.* **9** p 753–758
[14] Shimanskii A F, Podkopaev O I, Baranov V N 2015 *Adv Mat Res.* **1101** p 115-119
[15] Khirunenko L I, Pomozov Yu V, Sosnin M G et al. 2008 *Mater Sci Semicond Process* **11 (5)** p 344–347
[16] Cryse O, Vanhellomont J, Clawus P 2006 *Mater Sci Semicond Process* **9 (1)** p 246–251
[17] Bekman H H P, Gregorkiewicz T, Hidayat I F A et al. 1990 *Phys Rev B* **42 (16)** p 9802
[18] Inoue K, Taishi T, Tokumoto Y, Murao Y, Kutsukake K 2014 *Proceedings of the 12th Asia Pacific Physics Conference / JPS Conference Proceedings* **1 (1)**

IOP Conf. Series: Materials Science and Engineering **467** (2019) 012006 doi:10.1088/1757-899X/467/1/012006

About the limit state of deformable bodies

S Senashov[1] and I Savostyanova[2]

[1]Professor, Reshetnev Siberian State University of Science and Technology 660037, Krasnoyarsk, Russia
[2]Docent, Reshetnev Siberian State University of Science and Technology 660037, Krasnoyarsk, Russia

E-mail: sen@sibsau.ru

Abstract. The theory of limit state deals with statically determinate condition of solids. In this case the system is closed due to the limit conditions, such properties of matter as viscosity, elasticity, etc. cannot influence the limit state. In other words, being at the limit state the nature of the relationship between stress and strain has no effect on the limit state. The article discusses systems of equations which correspond to the classical theory of plasticity. It is assumed that the components of the velocity vector depend on two spatial coordinates only. The constructed system can be used to describe the torsion of a parallelepiped around the three orthogonal axes. For the constructed system of equations group point symmetries, conservation laws were found. It is shown that the system admits 8 -dimensional Lie algebra. On the basis of the symmetry group some classes of invariant solutions of rank 1 were constructed. They depend on the arbitrary functions of one variable. It is shown that these solutions can be used to describe plastic torsion of a parallelepiped around three orthogonal axes. It is shown that the system admits an infinite series of conservation laws.

1. Introduction

Some of the problems of the ideal plasticity theory are fairly well studied. These are the so-called statically determinate problems. Among these are the problems of the twisting of prismatic bars and the problems of the plane deformation state. They belong to a wide class of problems – the problems of the limit state of deformable bodies. The theory of the limit state is one of the fundamental sections of deformable solids mechanics [1]. The theory of the limit state deals with statically determinate state of solid bodies. In this case the system is closed due to the limit conditions, and such characteristics of the material as viscosity, elasticity etc. cannot influence the limit state. In other words when reaching the limit state, the nature of relations between strains and deformations does not influence the limit state. Some of these systems are described in [1].

In the proposed paper the system of equations of plasticity that describes the limit state is presented. This system can be used to govern the plastic flow around the three orthogonal axes.

2. Setting of a problem

Let us assume that $x = x_1, y = x_2, z = x_3$– orthogonal coordinates system, u, v, w – strain-rate vector components, e_{ij}- strain-rate tensor components, σ_{ij}- stress tensor components. The stress tensor components satisfy the equations of balance

$$\partial_j \sigma_{ij} = 0 \quad (1)$$

Over the repeated indices summation is assumed. The stress deviator and the strain-rate tensor are coaxial

$$\sigma_{ij} - \delta_{ij} p = \lambda e_{ij} \quad (2)$$

where σ_{ij}- Kronecker symbol, λ - any nonnegative function, $3p = \sigma_{ii}$.

System of equations (1)-(2) is closed by the von Mises yield criterion

$$(\sigma_{11} - p)^2 + (\sigma_{22} - p)^2 + (\sigma_{33} - p)^2 + 2(\sigma_{12}^2 + \sigma_{13}^2 + \sigma_{23}^2) = 2k_s^2 \quad (3)$$

It is known [1] that in case that the prismatic bar is twisted around axis oz then the strain-rate field is written as

$$u = -yz, v = xz, w = w(x, y) \quad (4)$$

When enlarging relations (1) we claim that

IOP Conf. Series: Materials Science and Engineering **467** (2019) 012006 doi:10.1088/1757-899X/467/1/012006

$$u = u(y,z), v = v(x,z), w = w(x,y) \tag{5}$$

Let us build up a system of equation corresponding to this strain-rate field. As a result we obtain the following system that will be studied in the paper

$$\partial_y \tau^1 + \partial_z \tau^2 = \partial_x p, \partial_x \tau^1 + \partial_z \tau^3 = \partial_y p, \partial_x \tau^2 + \partial_y \tau^3 = \partial_z p \tag{6}$$
$$(\tau^1)^2 + (\tau^2)^2 + (\tau^3)^2 = k_s^2,$$

where $\sigma_{12} = \tau^1, \sigma_{13} = \tau^2, \sigma_{13} = \tau^3$.

System of equations (6) can in particular be used to govern the twisting of an orthographical parallelelepiped in its plastic state around the three axes. See Fig. 1.

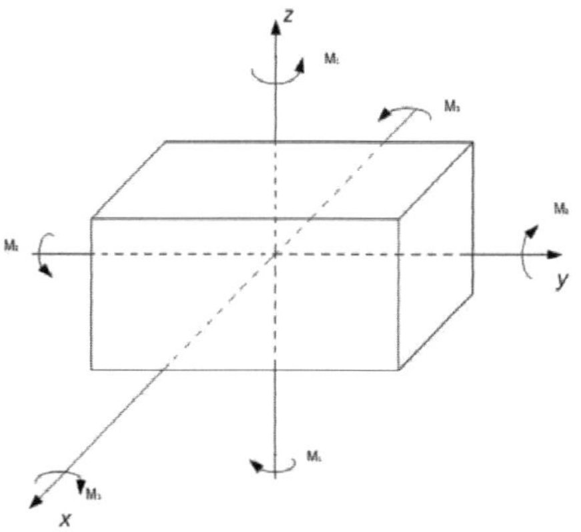

Figure 1. The torsion parallelepiped around the three axes

Let us assume that the parallelelepiped is twisted around axes ox, oy, oz with equal and opposite couples of forces with moments M_1, M_2, M_3. Wherein there are some limit moments M_1^*, M_2^*, M_3^*, when the parallelepiped goes into its plastic state and starts twisting. From system (6) we can see that such a problem is statically determinate and can be used to determine the value of the limit moments according to the formulas

$$M_1^* = \iint (y\tau^2 - z\tau^1) dydz, \tag{7}$$
$$M_2^* = \iint (z\tau^1 - x\tau^3) dxdz,$$
$$M_3^* = \iint (x\tau^3 - y\tau^2) dydz$$

Apart from the moments (7) hydrostatic pressure has effect on the body as well

$$P|_\Sigma = P_0,$$

where Σ - parallelepiped's lateral surface.

Let us study some of the attributes of the system (6).

3. Characteristic surfaces of System (6)

System (6) contains the finite relation that relates values τ^1, τ^2, τ^3. Let us differentiate it with respect to x,y,z and as a result we obtain the system

$$\partial_y \tau^1 + \partial_z \tau^2 = \partial_x p, \partial_x \tau^1 + \partial_z \tau^3 = \partial_y p, \partial_x \tau^2 + \partial_y \tau^3 = \partial_z p \tag{8}$$
$$\tau^1 \partial_x \tau^1 + \tau^2 \partial_x \tau^2 + \tau^3 \partial_x \tau^3 = 0,$$
$$\tau^1 \partial_y \tau^1 + \tau^2 \partial_y \tau^2 + \tau^3 \partial_y \tau^3 = 0,$$
$$\tau^1 \partial_z \tau^1 + \tau^2 \partial_z \tau^2 + \tau^3 \partial_z \tau^3 = 0.$$

Let us write the characteristic surface equation of system of equations (8) as

$$\psi = \psi(x,y,z) = 0 \tag{9}$$

The characteristic surfaces of system (8) are deduced from the determinant

$$\begin{vmatrix} \partial_x\psi & \partial_y\psi & \partial_z\psi & 0 \\ \partial_y\psi & \partial_x\psi & 0 & \partial_z\psi \\ \partial_z\psi & 0 & \partial_x\psi & \partial_y\psi \\ 0 & \tau^1 & \tau^2 & \tau^3 \end{vmatrix} = 0. \tag{10}$$

Note. It is not difficult to see that the last three equations of system (8) give the same rows in determinant (10) that is why they are omitted.

When expanding determinant (10) in the last row we obtain

$$\tau^1\partial_z\psi((\partial_z\psi)^2-(\partial_x\psi)^2-(\partial_y\psi)^2)+\tau^2\partial_y\psi((\partial_y\psi)^2-(\partial_x\psi)^2-(\partial_z\psi)^2)$$
$$+\tau^3\partial_x\psi((\partial_x\psi)^2-(\partial_y\psi)^2-(\partial_z\psi)^2)=0.$$

This equation can be written as

$$\tau^1 n_3(2n_3^2-1)+\tau^2 n_2(2n_2^2-1)+\tau^3 n_1(2n_1^2-1)=0, \tag{11}$$

where $n_1=\frac{\partial_x\psi}{\sqrt{(\nabla\psi)^2}}$, $n_2=\frac{\partial_y\psi}{\sqrt{(\nabla\psi)^2}}$, $n_3=\frac{\partial_z\psi}{\sqrt{(\nabla\psi)^2}}$. One of the solutions of equation (11) that does not depend on values τ^1,τ^2,τ^3 is

$$2n_i^2=1 \qquad i=1,2,3.$$

It follows that the angle between the normal and the characteristic surface $\psi(x,y,z)=0$ and vector n is$\pm{}^{\pi}/_{4}$. The set of the characteristic surface's elements create opening of cone $\pm{}^{\pi}/_{4}$around the direction that is determined by the third root of equation (11) and depends on the stress condition.

4. Point symmetries of System of equations (6)

Because system (6) contains the finite relation then it is necessary to work with its conclusion, which is written as (9), where the following notation is introduced $\partial_x\tau^1=q_1^1,\partial_x\tau^2=q_1^2,\partial_x\tau^3=q_1^3,\partial_x p=q_1^4\partial_y p=q_{2,}^4\partial_z p=q_3^4$.

$$\begin{gathered} q_2^1+q_3^2=q_1^4,\ q_1^1+q_3^3=q_2^4,q_1^2+q_2^3=q_3^4, \\ \tau^1q_1^1+\tau^2q_1^2+\tau^3q_1^3=0, \\ \tau^1q_2^1+\tau^2q_2^2+\tau^3q_2^3=0, \\ \tau^1q_3^1+\tau^2q_3^2+\tau^3q_3^3=0, \\ (\tau^1)^2+(\tau^2)^2+(\tau^3)^2=k_s^2. \end{gathered} \tag{12}$$

Let us search the point symmetries regarding which the variety determined by system of equations (12) is invariant.

According to Lie [2] procedure we search for the admissible operator of the point symmetry in the form of

$$X=\xi^j\frac{\partial}{\partial x_j}+\eta^i\frac{\partial}{\partial\tau^i}, \qquad j=1,2,3; i=1,2,3,4 \tag{13}$$

We extend operator (13) to the first-order derivatives using the formulas

$$\tilde{X}=X+\varsigma_k^i\frac{\partial}{\partial q_k^i}, \tag{14}$$

where $\varsigma_k^i=(\eta^i)-q_\beta^i D_k(\zeta^\beta), D_k=\frac{\partial}{\partial x_k}+q_k^i\frac{\partial}{\partial\tau^i}$. Operator (14) influences the system of equations (12) and we go to the manifold given by this system. As a result we obtain quadratic expressions with respect to «internal»- endogenous variables q_k^2,q_k^3. «External» - exogenous variables q_k^1,q_k^4 are determined from the system (12) via endogenous variables. In the obtained quadratic expressions we set the coefficients at the first and second degrees of the endogenous variables to zero. This allows us to obtain an overdetermined system of linear differential equations with respect to coefficients ξ^j,η^i. When solving this system we obtain the following result

Theorem. System of equations (6) admits Lie algebra L_8 , generated by the operators

$$X_i=\frac{\partial}{\partial x_i},X_4=x_i\frac{\partial}{\partial x_i},X_5=\frac{\partial}{\partial p},X_{12}=x_2\frac{\partial}{\partial x_1}-x_1\frac{\partial}{\partial x_2}+\tau^2\frac{\partial}{\partial\tau^1}-\tau^1\frac{\partial}{\partial\tau^2},$$
$$X_{13}=x_1\frac{\partial}{\partial x_3}-x_3\frac{\partial}{\partial x_1}+\tau^1\frac{\partial}{\partial\tau^3}-\tau^3\frac{\partial}{\partial\tau^1},X_{23}=x_3\frac{\partial}{\partial x_2}-x_2\frac{\partial}{\partial x_3}+\tau^3\frac{\partial}{\partial\tau^2}-\tau^2\frac{\partial}{\partial\tau^3}.$$

IOP Conf. Series: Materials Science and Engineering **467** (2019) 012006 doi:10.1088/1757-899X/467/1/012006

The presence of operators $X_i, i = 1,2,3,4$ means that system (6) allows x, y, z - direction translation and dilatation

$$x'_i = x_i + a_i, i = 1,2,3,\ x'_i = x_i exp a_4,$$

translation for hydrostatic pressure $p' = p + a_5$, and also rotation around the three coordinate axes, where a_i are group parameters.

5. Invariant solutions for System of equations (6)

1. Let us build an invariant solution with respect to the subalgebra generated by the operator $X_3 = \frac{\partial}{\partial z}$.
Such a solution should be determined in the form of

$$\tau^i = \tau^i(x,y), p = p(x,y). \tag{15}$$

Let us insert (16) into system (6) and we obtain

$$\partial_y \tau^1 = \partial_x p,\ \partial_x \tau^1 = \partial_y p,\ \partial_x \tau^2 + \partial_y \tau^3 = 0, \quad (\tau^1)^2 + (\tau^2)^2 + (\tau^3)^2 = k_s^2 \tag{16}$$

From (16) we can easily obtain

$$\tau^1 = f(x+y) + g(x-y), p = f(x+y) - g(x-y), \tag{17}$$

where $f(x+y), g(x-y)$ - some arbitrary differentiable functions.

Now functions τ^2, τ^3 will be determined from the system of equations

$$\partial_x \tau^2 + \partial_y \tau^3 = 0,\ (\tau^2)^2 + (\tau^3)^2 = k_s^2 - (\tau^1)^2. \tag{18}$$

System of equations (18) regulates the rotation of the bar in conditions when yield limit depends on variables *x,y*. Such problems are described in [1].

2. Let us build an invariant solution regarding the subalgebra generated by operator X_{12}. In the cylindrical coordinate system $r\theta z$ this operator is written as $X_{12} = \frac{\partial}{\partial \theta}$. In this case System (6) will be written as

$$\partial_\theta \tau_{r\theta} + \partial_z \tau_{rz} = r\partial_r p, \quad r\partial_r \tau_{r\theta} + r\partial_z \tau_{\theta z} + 2\tau_{r\theta} = \partial_\theta p, \quad r\partial_r \tau_{rz} + \partial_\theta \tau_{\theta z} + \tau_{rz} = \partial_z p, \quad \tau_{r\theta}^2 + \tau_{rz}^2 + \tau_{\theta z}^2 = k_s^2. \tag{19}$$

In this case the invariant solution is determined from the following system

$$\partial_z \tau_{rz} = r\partial_r p, \quad r\partial_r \tau_{r\theta} + r\partial_z \tau_{\theta z} + 2\tau_{r\theta} = 0, \quad r\partial_r \tau_{rz} + \tau_{rz} = \partial_z p, \quad \tau_{r\theta}^2 + \tau_{rz}^2 + \tau_{\theta z}^2 = k_s^2. \tag{20}$$

Then τ_{rz} is determined from the linear differential equation

$$r\partial_r \tau_{rz} + r^2 \partial_r^2 \tau_{rz} - \tau_{rz} + r^2 \partial_z^2 \tau_{rz} = 0.$$

The rest of the functions are determined from system (2).

6. Conservation laws of system of equations (6)

Let us determine the conservation laws of system of equations (6) in the form of
$\partial_x A(\tau^1,\tau^2,\tau^3,p) + \partial_y B(\tau^1,\tau^2,\tau^3,p) + \partial_z C(\tau^1,\tau^2,\tau^3,p) = 0.$

This equation should be valid for any solution of System of equations (6). From this the relations follow
$X_{12}A - \tau^2\partial_p B + \tau^1\partial_p C = 0,\ X_{13}B - \tau^3\partial_p A + \tau^1\partial_p C = 0,\ X_{12}B - \tau^2\partial_p A = 0, X_{12}C - \tau^1\partial_p A = 0,$
$X_{13}C + \tau^1\partial_p B = 0,\ X_{13}A - \tau^3\partial_p B = 0.$

Where $X_{12} = -\tau^2\partial_{\tau^1} + \tau^1\partial_{\tau^2}, X_{13} = -\tau^3\partial_{\tau^1} + \tau^1\partial_{\tau^3}$.

Let us show that these equations are simultaneous. Let $\partial_p A = \partial_p B = \partial_p C = 0$ then one of the solutions of this system is an infinitely series of the conservation laws

$$\partial_x A(S) + \partial_y B(S) + \partial_z C(S) = 0,$$

where $S = (\tau^1)^2 + (\tau^2)^2 + (\tau^3)^2,\ A(S), B(S), C(S)$_ some arbitrary differentiable functions. Conservation laws can be used to solve a number of boundary value problems. For the plasticity equations we did this in the papers [3 – 5].

Note. Are there any other laws? It is not known but in the author's opinion there are no other conservation laws.

7. About deformation fields of velocities of System (6).

System (6) allows us to find the stress condition. Let us assume that it is known. Then in order to find the three velocity vector components we have the three equations

$$2e_{12} = \partial_y u + \partial_x v, 2e_{13} = \partial_z u + \partial_x w, 2e_{23} = \partial_z v + \partial_y w, \tag{21}$$

where

$$2e_{12} = \partial_y u + \partial_x v, 2e_{13} = \partial_z u + \partial_x w, 2e_{23} = \partial_z v + \partial_y w. \tag{22}$$

Let us show that equations (21) can be solved in terms of the tensor component of the deformation velocities. As it is known, the deformation velocities tensor components apart from equations (21) must satisfy the compatibility equations as well. By virtue of relations (22) and (5) only six of them remain

$$\partial^2_{xy} e_{12} = 0, \partial^2_{xz} e_{13} = 0, \partial^2_{yz} e_{23} = 0, \tag{23}$$
$$\partial_x(\partial_x e_{23} - \partial_z e_{12} - \partial_y e_{13}) = 0,$$
$$\partial_y(\partial_y e_{13} - \partial_z e_{12} - \partial_x e_{23}) = 0,$$
$$\partial_z(\partial_z e_{12} - \partial_y e_{13} - \partial_x e_{23}) = 0.$$

Theorem. The compatibility equations of the deformation velocities are satisfied identically.

In this case from (21) we have

$$(\tau^1)^2(e_{12}^2 + e_{13}^2 + e_{23}^2) = k_s^2 e_{12}^2, \tag{24}$$
$$(\tau^2)^2(e_{12}^2 + e_{13}^2 + e_{23}^2) = k_s^2 e_{23}^2,$$
$$(\tau^3)^2(e_{12}^2 + e_{13}^2 + e_{23}^2) = k_s^2 e_{13}^2.$$

System of equations (24) is a system of homogeneous linear equations for variables $e_{12}^2, e_{13}^2, e_{23}^2$. Its determinant is written as

$$\begin{vmatrix} (\tau^1)^2 - k_s^2 & (\tau^1)^2 & (\tau^1)^2 \\ (\tau^2)^2 & (\tau^2)^2 - k_s^2 & (\tau^2)^2 \\ (\tau^3)^2 & (\tau^3)^2 & (\tau^3)^2 - k_s^2 \end{vmatrix}$$

This determinant is equal to zero because the sum of all the rows is equal to zero. This means that system (24) has only two independent equations for the three variables $e_{12}^2, e_{13}^2, e_{23}^2$. For example value e_{23}^2 can be selected randomly, that is why for the given stress condition found from system (6), the field of velocities is determined with a functional arbitrary rule.

8. Conclusion

The research performed has shown that the new system of equations describing the limiting state can be used to regulate the plastic flow around the three orthogonal axes. The system admits the eight-dimensional Lie algebra and an infinitely series of the conservation laws that allow building invariant solutions. The field of velocities for the given stress condition is determined with a functional arbitrary rule.

References

[1] Ivlev D D, Maksimova L A, Nepersin R I etc 2008 *The limit state of deformed bodies and moutainous rocks.* Moskow: Fizmatlit. chapter 1 pp 11-79 [in Russian]

[2] Polyanin A D , Zaitsev V F 2012 Handbook of nonlinear partial differential equation. NY:CRC Press, chapter 28, pp 1439 - 1457

[3] Senashov S,Yakhno A 2016 Application of conservation laws to Dirichlet problem for elliptic quasilinear systems *International Journal of Non-Linear Mechanics* **85** pp 1-5

[4] Senashov S, Cherepanova O, Kondrin A 2015 On Elastoplastic Torsion of a Rod with Multiply Connected Cross-Section *J. Siberian Federal Univ., Math. & Physics* **7(1)** p 343-351

[5] Senashov S., Cherepanova O., Kondrin A. 2014 Elastoplastic Bending of Beam. *J. Siberian Federal Univ., Math. & Physics* **7(2)** p 203-208

IOP Conf. Series: Materials Science and Engineering **467** (2019) 012007 doi:10.1088/1757-899X/467/1/012007

Modeling structural defect formation in cadmium telluride during electron irradiation

Yu. Yu. Loginov[1], A.V. Mozzherin[2], N.N. Paklin[2]

[1]Reshetnev Siberian State University of Science and Technology, 660037, Krasnoyarsk, Russia
[2]Siberian Federal University, 660041, Krasnoyarsk, Russia

E-mail: loginov@sibsau.ru

Abstract. We have conducted the formation of interstitial atoms and vacancies and made their clusters in the form of dislocation loops and voids in cadmium telluride under irradiation by electrons. We have calculated dependences of the radii of dislocation loops and voids depending on the irradiation time. We have compared the results of the simulation with the experimental data on the irradiation of CdTe in a transmission electron microscope as well.

1. Introduction

Modern technologies make it possible to produce semiconductor materials of very high quality with low content of structural defects, but the structural defects that remain in them can make serious changes in the physico-chemical properties, and, subsequently, affect the operation of devices based on them. Reliable operation of modern semiconductor devices of ultra-small sizes requires the control of the presence and density of not only traditional structural defects such as dislocations, stacking faults and micro-twins, but also defects of ≤ 10 nm in size. Devices have to work in unfavorable conditions, for example, under the influence of radiation with different energy. Such an impact on semiconductor materials causes the transformation and development of a defective network.

Cadmium telluride (CdTe) relating to compounds A_2B_6 is one of the main materials of semiconductor microelectronics. Cadmium telluride has a band gap of 1.5 eV, which provides an excellent match to the solar spectrum, a relatively high average atomic number, and the ability to achieve high resistivity values; it makes these materials ideal for gamma and x-ray detector devices [1]. However, the CdTe material is sensitive to effects of ionizing radiation because of the low value of stacking fault energy (SFE) ($\sim 11\pm 2$ mJ/m^2). Irradiation of CdTe by electrons in a transmission electron microscope causes similar evolution of defects [2]. We have studied the influence of electrons with sub-threshold energy on the processes of structural transformations in semiconductors in [3-7] where we have reported about the possibility of formation of point defects (vacancies and interstitials) in these materials under such irradiation. In relation to this, we have conducted a simulation of the processes of defect formation in CdTe under electron irradiation. Previously, we have developed chemical reaction rate in CdTe in order to describe the dynamics of electron irradiation damage in CdTe [8]. Taking into account the experimental studies reported in [7], we know that in addition to the formation of interstitial clusters, vacancy clusters are observed in CdTe, so the model is modified taking into account the formation of vacancy clusters and the results of the simulation are compared with the previously obtained experimental data.

2. Results and Discussion

Using effective values of activation energy and effective concentrations of interstitial atoms c_I and vacancies c_V, we have recorded equations simulating changes in concentrations c_I and c_V, as well as the growth of interstitial loops and pores with radii r_I and r_V and concentrations c_{bI} and c_{bv} in the following form:

$$\begin{cases} \frac{dc_I}{dt} = c_0 G - Kc_I c_V - 2K_{1I} c_I^2 - K_{1I} \frac{2\pi r_1}{b} c_I c_{bI} - K_{SI} c_I c_S, \\ \frac{dc_V}{dt} = c_0 G - Kc_I c_V - 2K_{1V} c_V^2 - K_{1V} \frac{4\pi r_V^2}{b^2} c_V c_{bV} - K_{SV} c_V c_S, \\ \frac{dc_{bI}}{dt} = c_0 K_{1I} c_I^2, \\ \frac{dc_{bV}}{dt} = c_0 K_{1V} c_V^2, \\ \frac{dr_I}{dt} = c_0 K_{1I} c_I b, \\ \frac{dr_V}{dt} = c_0 K_{1V} c_V b. \end{cases} \quad (1)$$

Here $c_0 = 1{,}5 \cdot 10^{22}$ cm^{-3} is the concentration of the nodes of the unirradiated crystal, $G = \sigma \cdot j$, $\sigma = 3 \cdot 10^{-22}$ cm^{2}is the electron scattering cross section at the lattice sites, $j = 10^{19}$ cm^{-2}c^{-1} is the electron flux density, $K_i = \nu \cdot \exp(-E_i / kT)$ is the reaction constant, $\nu$is - the frequency of vibration of atoms in the lattice, k is the Boltzmann constant, T is the temperature, E_i is the activation energy of processes, c_s is the concentration of "sinks" on the surface, $2\pi r_I / b$ and $4\pi r_V^2 / b^2$ is the number of seats joining point defects (PD) on the perimeter of the loop and on the void surface respectively, b is a parameter of the order of magnitude of the interatomic distance (Burgers vector). The value of $c_0 G$ is the generation rate of Frenkel pairs.

The recombination of the PD with the surface is performed as a result of the diffusion of the PD on the surface, the change in the concentration of the PD as a result of their displacement to the surface can be represented using the following formula:

$$\left.\frac{\partial c_{I,V}}{\partial t}\right|_{surface} = K_S c_{I,V} c_S = -D_{I,V} \frac{\partial^2 c_{I,V}}{\partial z^2}, \quad (2)$$

where z is the coordinate perpendicular to the sample surface and $D_{I,V} = b^2 \nu \cdot \exp(-E_{mI,mV} / kT)$ is the diffusion coefficient of interstitial atoms and vacancies. Taking (2) into account, it can be transformed (1) into a system of partial differential equations in the following way:

$$\begin{cases} \dot{u}_1 - D_I u_1'' - G = -c_0 K u_1 u_2 - 2c_0 K_I u_1^2 - \\ - 2\pi c_0 K_I u_1 u_3 u_5, \\ \dot{u}_2 - D_V u_2'' - G = -c_0 K u_1 u_2 - 2c_0 K_V u_2^2 - \\ - 4\pi c_0 K_V u_2 u_4 u_6^2, \\ \dot{u}_3 = c_0^2 K_I u_1^2, \\ \dot{u}_4 = c_0^2 K_V u_2^2, \\ \dot{u}_5 = c_0^2 K_I b u_1, \\ \dot{u}_6 = c_0^2 K_V b u_2, \end{cases} \quad (3)$$

where $u_1 = c_I / c_0$, $u_2 = c_V / c_0$, $u_3 = c_{bI} / c_0$, $u_4 = c_{bV} / c_0$, $u_5 = r_I / b$, $u_6 = r_b / b$. The dot and the prime denote partial derivatives with respect to the variables t and z, respectively.

The solution to the resulting system of equations (the numerical values of the parameters are given in [2]) gives the dependence u_1,..., u_6 on z and the irradiation time at 300 K (Fig. 1-3).

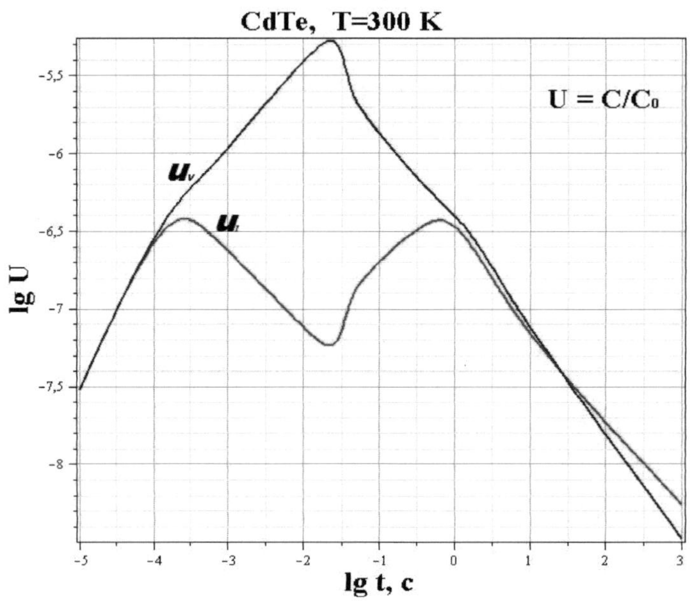

Figure 1. Dependence of u_I and u_V on the irradiation time at 300 K

The results presented in fig. 1 show that in the center of the sample (z = 100 nm) interstitial atoms have greater mobility than vacancies. When moving, the atoms quickly shift to the periphery and their concentration (c_I) decreases more intensively than the concentration of vacancies (c_V). In the time interval 0.1-0.01 seconds the concentration of vacancies reaches its maximum which practically coincides with the minimum of the concentration of interstitial atoms; it is explained by the dynamics of the processes of generation and recombination of Frenkel pairs, as well as agglomeration of Frenkel pairs in the di-interstitials and divacancies, which are assumed to be the nuclei for the formation of the dislocation loops and vacancy voids.

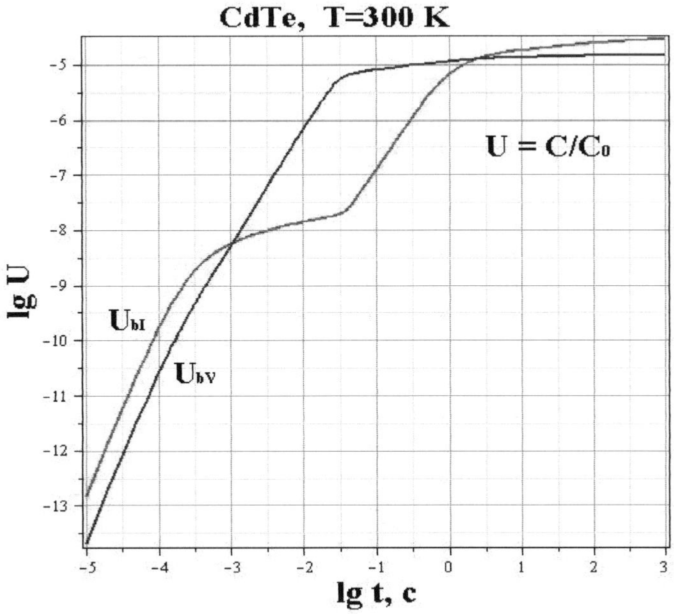

Figure 2. Change in the concentration of di-interstitials (u_3) and divacancies (u_4) within the irradiation time at 300 K

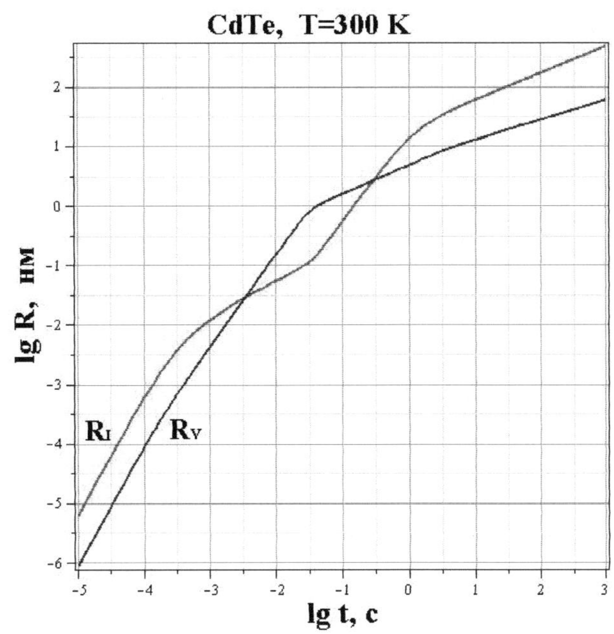

Figure 3. Dependence of the radii of the loops (u_5) and pores (u_6) on the irradiation time at 300 K

We note that the radii of dislocation loops and vacancy voids begin to exceed parameter b only a few seconds after the start of irradiation, and the final values of these radii practically do not depend on the initial conditions.

From Fig. 2 and 3, it can be seen that the growth of the radii of dislocation loops and the density of di-interstitials slow down. In the process of the sample irradiation the concentration of c_V begins to decrease, and c_I

IOP Conf. Series: Materials Science and Engineering **467** (2019) 012007 doi:10.1088/1757-899X/467/1/012007

increases, which causes the increase in the growth of the di-interstitials (u_3) and the loop radii (u_5) and slowing the growth of the divacancies (u_4) and the radii of voids (u_6).

3. Conclusion

Comparing the results of numerical modeling of cluster formation (point defects of interstitial and vacancy types) in cadmium telluride with the experimental data of a number of works [7,9], we come to the conclusion that the experiment and theory agree. However, in order to achieve maximum consistency, it is necessary to improve the existing model and introduce two parameters into it: firstly, taking into account the number of semiconductor nodes, since their number decreases upon irradiation; secondly, it is necessary to take into account the initial distribution of point defects, which is spatially inhomogeneous. The values of the initial concentrations depend on the energy of thermal activation of Frenkel pairs and the physical conditions under which the sample of the CdTe crystal is processed. This will allow to confirm the initial conditions of the experiment and theoretical modeling more accurately.

References

[1] Elsharkawy M, Kanda G, Abdel-Hady E, Keeble D 2016 *Appl. Phys. Lett.*, **108** 242102
[2] Loginov Y, Brown P, Durose K*The Regularities of Structural Defect Formation in Semiconductors II–VI* (Logos, Moscow, 2003)
[3] Varley J 1954 Nature, 174 886
[4] Vavilov V, Kiv A, Niyazova O 1975 *Phys. Stat. Sol.(a)* 32 11
[5] Yunusov M, Zaikovskaya M, Oksengendler B, Tokhirov K 1976 *Phys. Stat. Sol.(a)* 35 K145
[6] Karpov V and Klinger M 1987 *Phys. Tech. Sem.* 12 1887
[7] Loginov Yu and Brown P 1992 *Phys. Stat. Sol. (a)* 132 323
[8] Gué A, Djafari-Rouhani M, Estéve D, IdrissiSaba H 1991 *J. Phys. I*, **1** 97
[9] Loginov Y, Mozzherin A, Brilkov A 2014 *Phys. Solid State***56** 720

IOP Conf. Series: Materials Science and Engineering **467** (2019) 012008 doi:10.1088/1757-899X/467/1/012008

Comparative analysis of electronic loads with two regulators

E. A. Mizrah, D. K. Lobanov, A. S. Fedchenko, D. N. Poymanov

Reshetnev Siberian State University of Science and Technology, 660037, Krasnoyarsk, Russia

E-mail: enis-home@mail.ru

Abstract. While developing and testing spacecraft power supply systems, we need devices that simulate various electrical characteristics of the spacecraft payload – electronic loads. Providing high bandwidth of these electronic loads at power carrying several kW may become a difficult task, and one of possible solutions is the use of electronic loads with two controllable regulators – a linear regulator and a switching regulator. Such electronic loads allow combining wide bandwidth of a linear regulator and possibility of energy recovery provided by a switched regulator. In the paper we studied possible topologies of these electronic loads, we developed a mathematical description of the electronic load used in the test, we performed the analysis of admittances and quality of noise suppression for various topologies. The most promising topologies that can be used as electronic loads for spacecraft power systems tests were highlighted.

1. Introduction

Electrical testing of spacecraft power supply system (SPS) is an essential part of spacecraft development process. Compliance of the developed SPS with the claimed requirements is determined during these tests, which are carried out using specialized electronic loads (EL) that simulate static and dynamic electrical characteristics of the payload. Since these ELs are used to solve a wide range of tasks arising during testing [1-6], a number of technical requirements to the load are specified:

- ranges of input voltage and current;
- shapes of input I-V curves;
- permissible deviations of stabilized quantities;
- energy utilization factor (when energy recovery is used);
- input current bandwidth (when EL is used as an AC current source).

Providing all the specified requirements simultaneously may become a difficult task, especially if it is necessary to provide high bandwidth of the input current stabilization at a power carrying several kW. One of the possible solutions is the use of electronic loads with two controllable regulators: a linear regulator and a switching regulator. These ELs allow combining the wide bandwidth of the linear regulator and the possibility of energy recovery (provided by a switched regulator), which also leads to the improvement of mass-dimensional characteristics of EL. [7-15]

2. Possible structures of electronic loads with two regulators.

The main idea of EL with two control loops is that the stabilization of input current is performed by a linear regulator (LR), while a switching regulator (SR) limits the power dissipated by LR. There are two possible ways of connecting LR and SR in such EL: parallel and series. In a parallel circuit the power stabilization of LR is achieved by limiting the LR current, in series – by limiting the LR voltage. In other words, SR in the parallel circuit performs the function of a current stabilizer, and in the serial circuit it serves as a voltage stabilizer. Control loops in the parallel circuit must contain two current sensors (one is for LR and another one is for SR), in the series circuit – one current sensor (CS) and one voltage sensor (VS). In order to simulate different types of input I-V curves (constant power and/or constant resistance stabilization), it is also necessary to add an input voltage sensor to both circuits. There are three possible locations for the current sensors in the parallel circuit and two possible locations in the series circuit; that leads to eight possible different topologies of EL with two control loops (figure 1). Table 1 gives general comparison of parallel and series topologies.

Table 1. Common features of parallel and series topologies

	Parallel topologies (a)-(f)	Series topologies (g)-(h)
Advantage	Possibility of independent current stabilization through LR and SR	Wide-bandwidth LR suppresses disturbances in input current more effectively
Disadvantage	An amplitude of induced pulsations is limited by the magnitude of the LR current	Performance of circuit strongly depends on input line/source impedance

In figure 1, the following designations are used for the set-points: I_{in} – required EL input current, I_{LR} – required LR current, I_{LRmax} – maximum permissible LR current, V_{LR} – required LR voltage, $V_{ип}$ – required SR voltage. Reference signals of LR and SR are generated by the control unit (CU) and depend on the position of the current sensors. In addition, CU can perform protection functions; it allows to reduce the total cost of the EL. However, in topologies (b)-(e) and (g)-(h), CU receives information about the magnitude of the current only in one of the elements of EL (LR or SR) which creates additional difficulties in the case when CU performs as a protection device as well.

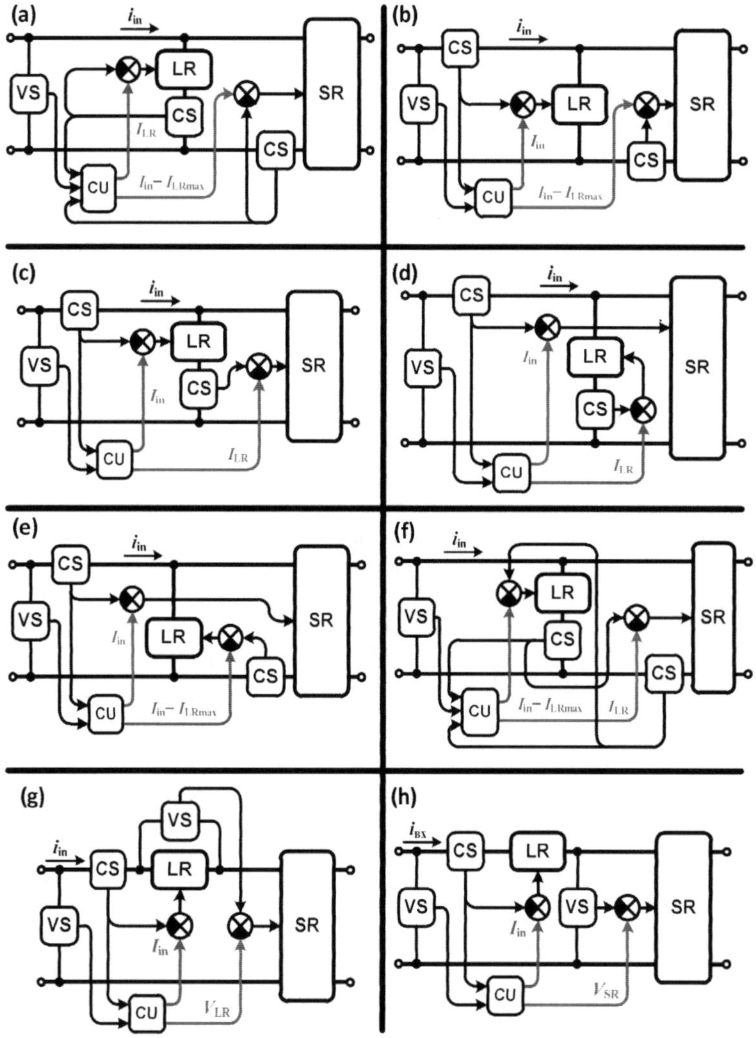

Figure 1. Topologies of electronic loads with two control loops

Table 2. Individual features of each topology

	(a)	(b)	(c)	(d)	(e)	(f)	(g)	(h)
1. Increased durability	+	+		+				
2. Independent interference induction by LR/SR	+		+				+	
3. Wide bandwidth of inducted test signal	+	+	+					
4. LR and SR can be developed independently	+	+		+				
5. Input current filtration		+	+				+	+
6. Input current magnitude monitoring		+	+	+	+		+	+
7. Diagnostics of LR and SR	+					+		
8. Input current can be increased rapidly	+	+	+	+	+	+		
9. Input current can be decreased rapidly							+	+

Table 2 shows distinguishing features of each topology, namely:

1) Increased durability – if LR breaks down, there is an opportunity to continue the work by means of SR only;

2) Independent interference induction by LR/SR – the possibility to induce noise of arbitrary waveform on LR or SR current independently; it allows achieving minimal harmonic distortion of this waveform;

3) Wide bandwidth of inducted test signal – noise induction is performed through LR which has wide bandwidth;

4) LR and SR can be developed independently – LR or SR can be turned on and tested; it is not necessary to connect it to the second regulator (SR or LR respectively);

5) Input current filtration – LR reduces noise and pulsations induced on EL input current, in particular, LR reduces input current ripples of SR;

6) Input current magnitude monitoring – topology contains current sensor that measures the main stabilizing variable – EL input current; it leads to increasing of stabilization accuracy;

7) Diagnostics of LR and SR – it is possible to implement the diagnostics function in the control unit without introducing additional elements into the circuit;

8) Input current can be increased rapidly – in parallel topologies (a)-(f), the current increases with the rate corresponding to the bandwidth of LR (but when the current decreases significantly, there is a risk that LR can close completely, and then this current continues to decrease with the rate corresponding to the bandwidth of SR, which is significantly smaller than the bandwidth of LR);

9) Input current can be decreased rapidly – in series topologies (g)-(h), the current decreases with the rate corresponding to the bandwidth of LR (but when the current increases significantly, there is a risk that LR can open completely, and then this current continues to increase with the rate corresponding to the bandwidth of SR).

On the basis of the above mentioned positive and negative qualities of each topology, we concluded that the topologies (a)-(c) and (g) were of the greatest interest for further consideration.

3. Mathematical model

Figure 2 shows the testing circuit for SPSS with EL that is described by the following system of equations:

$$\begin{cases} v_{\text{in}} = e_{\text{SPS}} - i_{\text{in}} Z_{\text{SPS}} \\ i_{\text{out}} = \left(v_{\text{SRout}} - e_{\text{DC}} \right) Y_{\text{DC}} \end{cases}, \tag{1}$$

where v_{in} – EL input voltage, e_{SPS} – equivalent output voltage of SPS, i_{in} – EL input current, Z_{SPS} – equivalent common impedance of SPS, i_{out} – EL output current (recovered current), v_{SRout} – EL output voltage, e_{DC} – equivalent voltage of DC network (to which energy recovery is performed), Y_{DC} – equivalent common admittance of DC network.

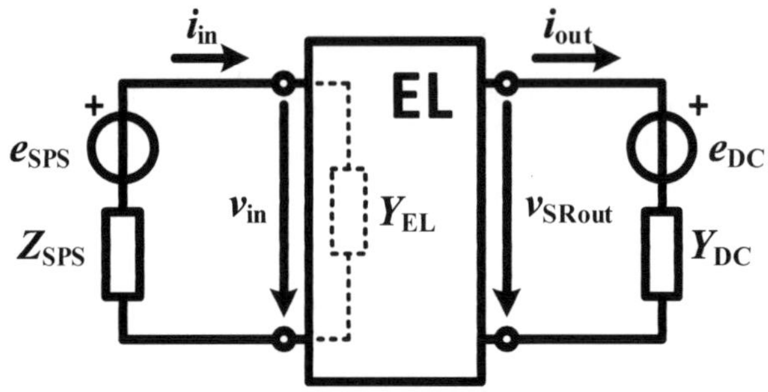

Figure 2. EL as a part of testing setup

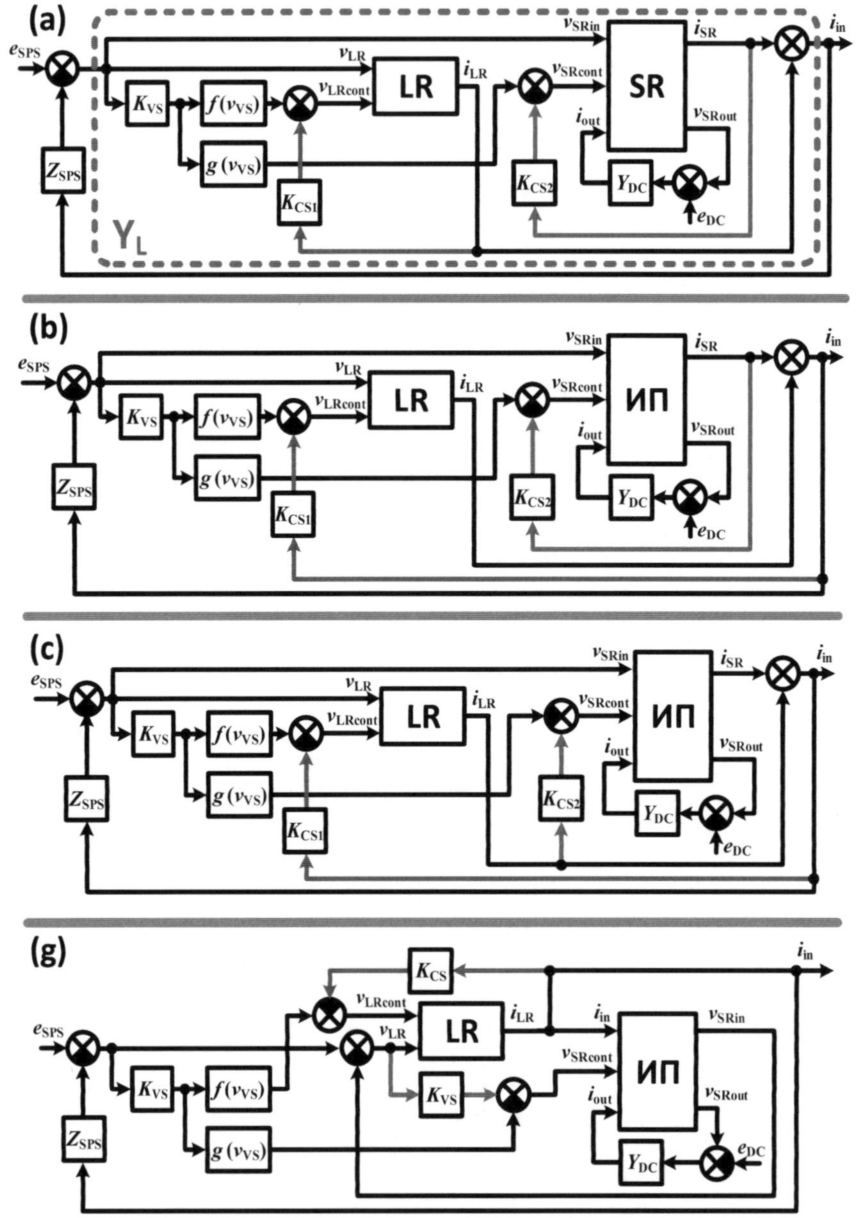

Figure 3. Topologies of electronic loads with two control loops

A control unit (CU) in general case is described by nonlinear dependences of the control voltage of LR (v_{LRcont}) or SR (v_{SRcont}) on the voltage of sensor input v_{VS}. Further, these dependencies will be denoted as $f(K_{VS}v_{in})$ for LR and $g(K_{VS}v_{in})$ for SR, where K_{VS} is voltage sensor gain.

On the basis of the functional relationships presented in figure 1, it is possible to compile systems of equations describing the relationships between current and voltage in the circuits. The topology (a) is described as:

$$\begin{cases} v_{LRcont} = f(K_{VS}v_{LR}) - i_{LR}K_{CS1} \\ v_{SRcont} = g(K_{VS}v_{LR}) - i_{SR}K_{CS2} \\ i_{in} = i_{LR} + i_{SR} \end{cases}, \quad (2)$$

where K_{CS1} and K_{CS2} are current sensors gains, i_{LR} and i_{SR} – LR and SR current respectively.

The topology (b) is described as:

$$\begin{cases} v_{LRcont} = f(K_{VS}v_{LR}) - i_{in}K_{CS1} \\ v_{SRcont} = g(K_{VS}v_{LR}) - i_{SR}K_{CS2} \\ i_{in} = i_{LR} + i_{SR} \end{cases}. \quad (3)$$

The topology (c) is described as:

$$\begin{cases} v_{LRcont} = f(K_{VS}v_{LR}) - i_{in}K_{CS1} \\ v_{SRcont} = i_{LR}K_{CS2} - g(K_{VS}v_{LR}) \\ i_{in} = i_{LR} + i_{SR} \end{cases}. \quad (4)$$

The topology (d) is described as:

$$\begin{cases} v_{LRcont} = f(K_{VS}v_{LR}) - i_{in}K_{CS} \\ v_{SRcont} = v_{LR}K_{VS} - g(K_{VS}v_{LR}) \\ v_{in} = v_{LR} + v_{SR} \end{cases}. \quad (5)$$

On the basis of systems (1)-(5) we constructed the functional diagrams shown in figure 3.

The description of the "LR" block can be obtained both analytically (based on the system of equations describing the structure and elements of the LR) and by identification of the real LR. Figure 4 (a) shows the internal structure of LR on the basis of an identified model in two inputs and single output (MISO), where: $W_{C1}(s)$ is the transfer function of the LR controller, V_F – forward voltage of p-n junction of controlled switch, $\mathbf{X}_1$, $\mathbf{U}_1 = [v_{LR}\ v_{LRcont}]^T$, $\mathbf{Y}_1 = [i_{LR}]$ – state, control and output vectors respectively, $\mathbf{A}_1$, $\mathbf{B}_1$, $\mathbf{C}_1$, $\mathbf{D}_1$ – coefficient matrices of the identified LR model. The scheme of identified MISO-model being used is shown in figure 4 (b).

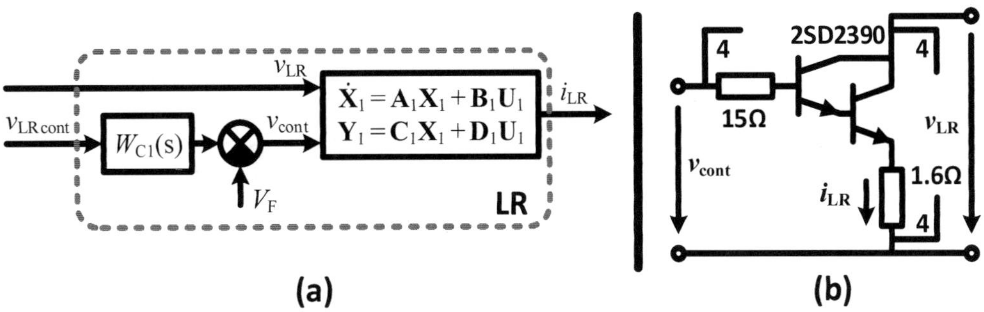

Figure 4. Examined linear regulator model: (a) block diagram, (b) electrical circuit

Identification of the SR model in the "SR" block in figure 5(a) is difficult due to nonlinear operation of SR. In cases where the cutoff frequency of the SR is significantly lower than its switching frequency, SR can be represented as a continuous linear model. Using of continuous linear model of SR is quite justified, because in electronic loads there are no bandwidth requirements that are imposed on SR. In addition to the MIMO-model, the "SR" block also includes $W_{C2}(s)$ – transfer function of SR gain and $K_{PWM} = 1/V_{saw}$ – the pulse width modulation (PWM) gain, where V_{saw} is the magnitude of PWM sawtooth.

It is possible to use any switch-mode converter as SR that meets the requirements for the device. However, in order to protect the equipment under test, there is a requirement for the galvanic isolation of the input and output terminals, which can be realized by means of bridge converters based on the boost and buck schemes (figure 5). Due to the fact that at the input of SR it is necessary to maintain constant current, we have to set a filter at the input of SR that will smooth the current (inductor) or voltage (capacitor) ripples.

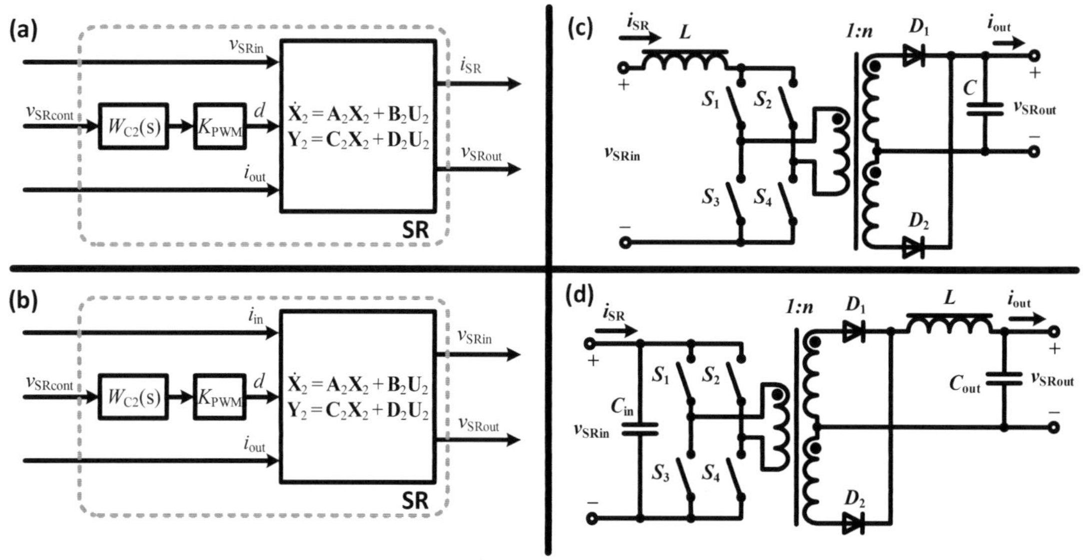

Figure 5. Examined switched-mode converter models: (a) block diagram and (б) circuit of boost-based full-bridge converter, (в) and (г) – of buck-based full-bridge converter

It should be noted that in the buck-based circuit (figures 5(c) and 5(d)), when large input current is stabilized, the pulsed current flowing through the input capacitance C_{in} becomes significant. This current can lead to heating of the input capacitor and, as a consequence, to the failure of SR. Thus, in parallel topologies (a)-(c), it is more preferable to use boost-derived converters (figures 5(a) and 5(b)), which are described by the following system of nonlinear equations:

$$\begin{cases} L\dfrac{di_{SR}}{dt} = v_{SRin} - \dfrac{2}{n}v_{SRout}\cdot(1-d) \\ C\dfrac{dv_{SRin}}{dt} = \dfrac{2}{n}i_{SR}\cdot(1-d) - i_{out} \end{cases}, \quad (6)$$

where i_{SR} – SR input current, v_{SRin} – SR input voltage, v_{SRout} – SR output voltage, d – duty ratio, i_{out} – SR output current (recovery current), $n = N_2/N_1$ –ratio of turns of a power transformer (N_1 и N_2 – number of turns on the primary and secondary windings respectively), L - inductance, C - capacitance.

In the series circuit (g) the change in the value of the EL input current leads to the change of LR voltage. The rate of this change is much greater than can be compensated by the SR because when the high-frequency interference is induced, the current amplitude is limited by active and inductive components of the impedance of the cable line and the input impedance of SR. In order to increase the amplitude of induced interference, it is necessary to reduce the value of the input impedance of the SR at high frequencies which can be achieved by increasing the capacitance of the SR input capacitor. Thus, in the series topology (g) it is recommended to use buck-derived bridge converters (figures 5(c) and 5(d)) which are described by the following system of nonlinear equations:

IOP Conf. Series: Materials Science and Engineering **467** (2019) 012008 doi:10.1088/1757-899X/467/1/012008

$$\begin{cases} C_{\text{in}} \dfrac{dv_{\text{SRin}}}{dt} = i_{\text{SR}} - 2 \cdot n \cdot i_{\text{L}} \cdot d \\ L \dfrac{di_{\text{L}}}{dt} = 2 \cdot n \cdot v_{\text{SRin}} \cdot d - v_{\text{SRout}}, \\ C_{\text{out}} \dfrac{dv_{\text{SRout}}}{dt} = i_{\text{L}} - i_{\text{out}} \end{cases} \tag{7}$$

where C_{in} and C_{out} – capacities of input and output capacitors respectively, i_{L} – inductor current.

4. Simulation results

For the further research, four EL topologies (a)-(c) and (g) were developed, each of them meets the following requirements:

- input current – 10 A;
- mean power dissipated by the LR – 100 W;
- the largest deviation of the input current from the stabilized value – 0.1 A;
- the largest peak-to-peak voltage of the input current pulsations – 1 A;
- bandwidth – not less than 100 kHz.

Admittance response. One of the conditions for ensuring stability is the fulfillment of the inequality $Z_{\text{SPS}}Y_{\text{L}} < 1$, where Z_{SPS} – the SPS output impedance, $Y_{\text{L}} = I_{\text{in}}/V_{\text{in}}$ – input admittance of EL (figure 3(a)). For a current source, the best situation is when the conductivity over the entire frequency range is equal to zero.

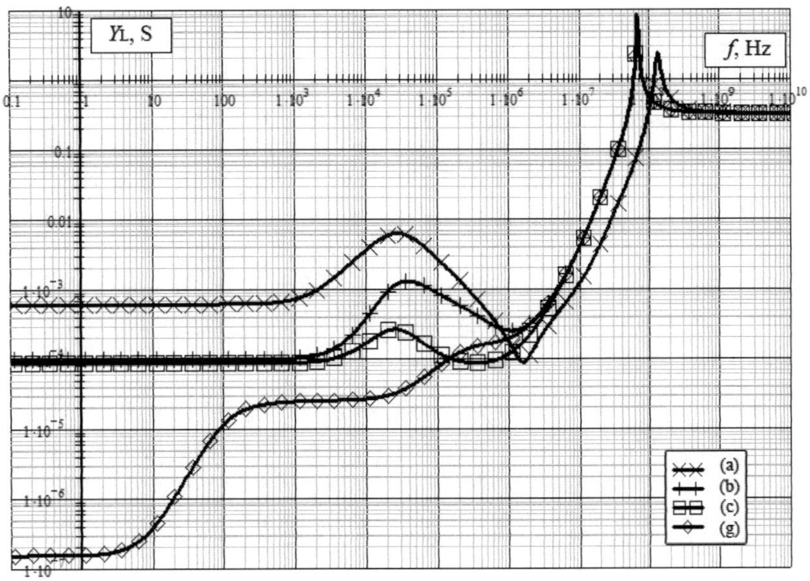

Figure 6. Admittance response of examined topologies

From the admittance responses (figure 6) it can be can concluded that the parallel topologies (b) and (c) have approximately equal input admittance practically on the entire frequency range, while the topology (a) has much higher admittance at the low frequency range. The topology (g) has the best admittance response for the current source which is equal only to tenths of a μS at low frequencies.

Noise interference. The transfer functions of closed control loops are defined as $W(s) = I_{\text{in}}(s)/V_{\text{ref}}(s)$ (figure 7). In topologies (a) and (b) $V_{\text{ref}}(s)$ is the reference voltage of both LR and SR, in topologies (c) and (d) V_{ref} is the reference voltage only for LR.

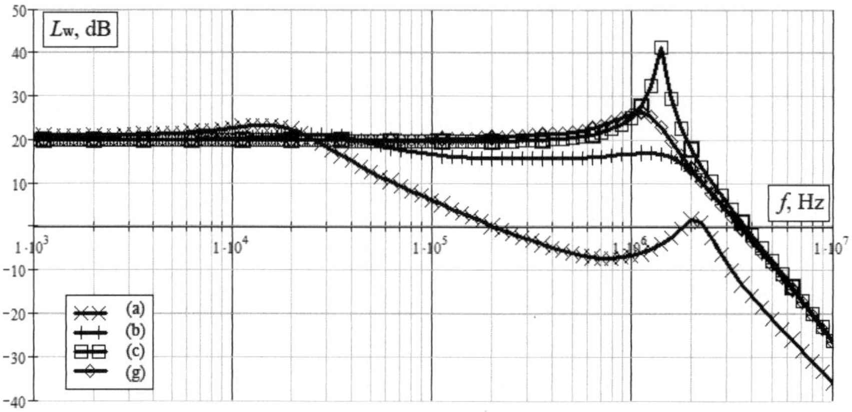

Figure 7. Closed loop transfer functions of examined topologies

There is a decay beginning at approximately 10 kHz in topology (a) which is caused by cut-off frequency of SR. In topology (b), unlike the topologies (c) and (g), there is no explicit resonant peak, which makes it possible to impose test signals without distortion having more complex harmonic composition.

5. Discussion and conclusion

The obtained results make it possible to draw a conclusion regarding the choice of a particular topology on the basis of the EL requirements. With more stringent requirements for survivability, the use of topologies (a) and (b) is more appropriate since they can provide continuation of work in cases where LR is in emergency mode. Topologies (a) and (b) are also simpler in design and development as it is possible to design LR and SR independently.

If it is necessary to induce the pulsations of large amplitude while maintaining high recovery rate, the series topology (g) is the most suitable. However, topologies (b) and (c) have wider bandwidth; it allows to induce pulsations on input current of higher frequency. The topology (g) is characterized by the smallest values of the input admittance at low frequencies, but, on the other hand, the topology (g) is more sensitive to the value of output inductance of the tested SPS.

Acknowledgements

This study was supported by the Ministry of Education and Science of the Russian Federation (Government Contract 14.577.21.0082, unique identifier RFMEFI57714X0082).

References

[1] Duran E., Andujar J.M., Segura F., Barragan A.J. A high-flexibility DC load for fuel cell and solar arrays power sources based on DC–DC converters. Applied Energy Journal, 2011. Vol. 88. P. 1690–1702. doi: 10.1016/j.apenergy.2010.11.002.

[2] Locment F., Sechilariu M., Houssamo I. DC Load and Batteries Control Limitations for Photovoltaic Systems. Experimental Validation. IEEE Trans. on Power Electronics, 2012. Vol. 27(9). P. 4030–4038. doi: 10.1109/TPEL.2012.2189134.

[3] Ceylan M., Balikci A. Design and Implementation of an Electronic Constant Current DC Load for Battery Discharge and Power Supply Test Systems. 16th International Power Electronics and Motion Control Conference and Exposition, 2014. P. 924–927. doi: 10.1109/EPEPEMC.2014.6980625.

[4] Locment F., Sechilariu M., Houssamo I. A Sliding-Mode Duty-Ratio Controller for DC/DC Buck Converters With Constant Power Loads. IEEE Trans. on Power Electronics, 2012. Vol. 27(9). P. 4030–4038. doi: 10.1109/TPEL.2012.2189134.

[5] Kakigano H., Miura Y., Ise T. Low-Voltage Bipolar-Type DC Microgrid for Super High Quality Distribution. IEEE Trans. on Power Electronics, 2010. Vol. 25(12). P. 3066–3075. doi: 10.1109/TPEL.2010.2077682.

[6] Wang S., Li M., Chen Z. et al. Design and Implementation of Power Electronic Load Used to Test Tidal Current Energy Generator Sets. IEEE International Conference on Fuzzy Systems (FUZZ-IEEE), 2014. P. 354-358. doi: 10.1109/FUZZ-IEEE.2014.6891752.

[7] Denk F., Haehre K., Koerner J. et al. Adjustable 20 kW full-SiC electronic load with energy recovery for

medium-frequency inverter. PCIM Europe 2016; International Exhibition and Conference for Power Electronics, Intelligent Motion, Renewable Energy and Energy Management; Proceedings of, 2016. P. 1721–1727.

[8] Bouaicha A., Allagui H., Mami A. Study of an electronic load for measuring the internal impedance of a PEM fuel cell. Systems, Signals & Devices (SSD), 2013 10th International Multi-Conference on, 2013. P. 1–5.

[9] Ibrahim O., Yahaya N.Z., Saad N. et al. Design and Analysis of a Digital Controller for Boost Converter with Renewable Energy Sources for Domestic DC Load. Applied Mechanics and Materials, 2015. Vol. 785. P. 141-145. doi: 10.4028/www.scientific.net/AMM.785.141.

[10] Hu G., Wie Y., Lei H. et al. Constant Current Control of DC Electronic Load based on Boost Topology. Electronics and Electrical Engineering, 2014. Vol. 20(2). P. 36-39. doi: 10.5755/j01.eee.20.2.6381.

[11] Ying C. Design of 300A constant current electronic load. Selected Papers of the Photoelectronic Technology Committee Conferences, 2015. P. 1-7. doi: 10.1117/12.2228515.

[12] She X., Zou Y., Wang C. et al. Research on Power Electronic Load: Topology, Modeling, and Control. Applied Power Electronics Conference and Exposition, 2009. APEC 2009. Twenty-Fourth Annual IEEE, 2009. P. 1661-1666. doi: 10.1109/APEC.2009.4802891.

[13] Tsang K.M., Chan W.L. Fast Acting Regenerative DC Electronic Load Based on a SEPIC Converter. IEEE Trans. on Power Electronics, 2012. Vol. 27(1). P. 269-275. doi: 10.1109/TPEL.2011.2158446.

[14] Upadhyay S., Mishra S., Joshi A. A Wide Bandwidth Electronic Load. IEEE Trans. on Industrial Electronics. 2012. Vol. 59(2). P. 733-739. doi: 10.1109/TIE.2011.2148680.

[15] Fedchenko A.S., Lobanov D.K., Mizrah E.A. Design Principles and Classification of the Adjustable Electronic Loads of Electrical Power Systems of Spacecrafts. International Journal of Applied Engineering Research (IJAER), 2015. Vol. 10(20). P. 41004-41010.

IOP Conf. Series: Materials Science and Engineering **467** (2019) 012009 doi:10.1088/1757-899X/467/1/012009

Seam tracking during electron beam welding in the air

V. Braverman and V. Bogdanov

Reshetnev Siberian State University of Science and Technology, 660037, Krasnoyarsk, Russia

E-mail: braverman-vladimir@rambler.ru

Abstract. We study methods of beam positioning at the joint during Electron Beam Welding in the air. Traditional methods of automatic beam positioning at the joint are unacceptable during the Electron Beam Welding in the air because of the significant dispersion of electrons and the impossibility of the beam deflection inside the electron beam gun due to the presence of an airlock. For joint tracking, we propose using magnetic fields of the current in the welded parts created by the beam current. It is established that the vertical component of the magnetic field of the current in the welded parts is proportional to the beam deflection at the joint. A differential flux gate meter is used as a tracking device. We outline the functional diagram of the joint tracking device and address issues with error prevention methods.

Introduction. Recently, the industrial use of Electron Beam Welding in the air has grown significantly. However, issues with precise positioning of the electron beam at the joint of the welded parts are the same as in the Electron Beam Welding in the vacuum.

Traditional methods of automatic beam positioning at the joint are unacceptable during the Electron Beam Welding in the air because of the significant dispersion of electrons and the impossibility of the beam deflection inside the electron beam gun due to the presence of an airlock. For seam tracking we propose using magnetic fields of the current in the welded parts created by the beam current. It is established that the vertical component of the magnetic field of the current in the welded parts is proportionate to the beam deflection along the seam. A differential flux gate meter is used as a tracking device. We outline the functional diagram is of the seam tracking device and address issues with error prevention methods.

Several aspects of this subject are covered in this work [1]. We describe a method of determining the beam position along the seam and provide a functional diagram of the device to help apply this method in the Electron Beam Welding in the air.

The method of determining the beam position at the joint. The proposed method is based on identification of a magnetic field of the welding current (I_B) and the beam coordinates. The main idea of this method is that when a beam deflection from the joint occurs, a redistribution of welding current components and the current-induced magnetic fields follow [2], [3].

The electron beam current I_B is divided into two components I_1 and I_2 (Fig. 1) with the help of current collectors. When the electron beam is located right above the joint then the magnetic-fields strengths H_1 and H_2 have equal values and opposite directions (Fig. 1, *a*). In this case, the resulting magnetic field is defined by current I_B, and the vector of the magnetic-field strength is located in the horizontal plane. With the beam deflection (-ε) from the joint (Fig 1, *b*), the current I_2 flows from the beam to the current collector through the welded part, and the magnetic-field strength produced by the current I_2 is changed by the value of H_V. The resulting magnetic field over the welded part of the joint is defined by the vertical component of the magnetic field strength H_V from the current I_2, flowing through the welded part of the joint, and the horizontal component of magnetic field strength from the beam current I_B.

When the beam deflection is in the opposite direction, the current I_1 flows through the welded part of the joint and the magnetic field strength produced by current I_1 is changed by the value of (-H_V) (Fig. 1, *c*). The resulting magnetic field over the welded part of the joint is defined by the vertical component of the magnetic field strength (-H_V) from the current I_1, flowing through the welded part of the joint, and the horizontal component of magnetic field strength from the beam current I_B. Therefore, the vertical component of the magnetic field over the welded part of the joint carries the information about the beam position at the joint. If a magnetic field sensor (for example, a fluxgate meter) is used in such a way that its sensitive axis is vertically placed, then it is possible to measure the vertical components (H_V) and (-H_V), which carry the information about the beam position at the joint.

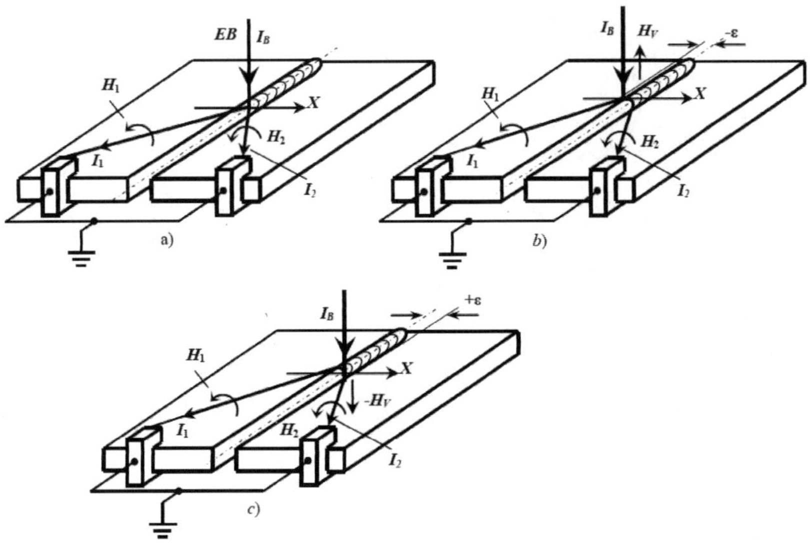

Figure 1. Dependence of the magnetic field vertical component on the electron beam position: a) – X=ε=0 (no displacement); b - X=-ε (left displacement); c – X=+ε (right displacement)

Theoretical analysis. The analytical definition of the magnetic field strength dependence on the electron beam deflection from the joint is carried out with a 3*D* model, which simulates the welding process based on the electromagnetic properties of air restricted in the plane of the welded parts with coordinates determined by the two-dimensional model analysis. This dependence is defined by the following equation:

$$H_V = F(x, y, h_д, I),$$

where H_V is a vertical component of the magnetic field; x, y are beam coordinates; h_S is the height of the flux-gate meter position; I is welding current.

The simulation is based on Maxwell equations calculation [1]. Two parameters were modeled – the direct current in conducting medium and the electromagnetic field. The equations were solved with the finite elements method. The calculations were produced for aluminum alloy and steel materials under beam deflection variations.

The calculation results for the process of welding the parts made of aluminum alloy with thickness of 10 mm with the I_B = 250 *mA* and the welding process speed of S_W = 0,5 *sm/s* are presented in Fig. 2. The lines represent currents and the shaded areas represent the magnetic field distribution.

This shows that when there is no beam deflection, the distribution of welding current components I_1 and I_2 and the magnetic fields produced by them are symmetrical relative to the joint (Fig. 2, *a*). When there is deflection from the joint, the symmetry is disturbed (fig 2, *b*), and some current created by the part with the beam deflection passes through the welding area. The larger the beam deflection is, the higher the symmetry disturbance is. The same effect is observed in the simulation described above.

As a result, the parity of the currents I_1 and I_2 no longer exists and the vertical component of the magnetic field H_V appears. The value of H_V depends on the value and the direction of the beam deflection from the joint (Fig 2). The calculation results have shown that beam deflection from the joint is converted to deviation of vertical component of the magnetic field.

The calculations have shown that magnetic field strength H_V is almost proportionate to the beam deflection from the joint. The research results provide evidence that it is possible to determine the beam position relative to the joint by the value and direction of the vertical component of the magnetic fields. The sensor installation at some distance from the beam does not lead to a systematic error because the vertical component H_V appears when there is a beam deflection form the joint.

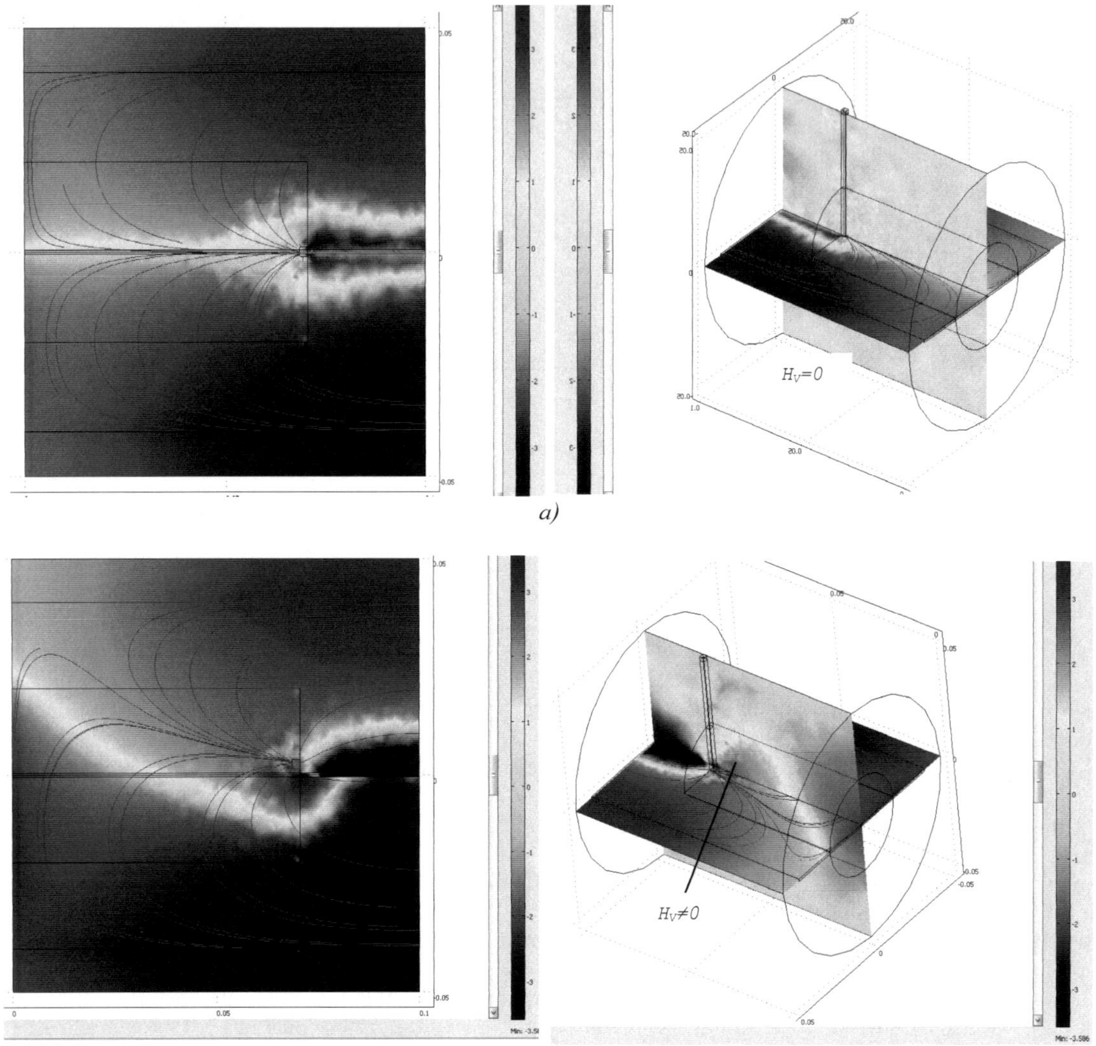

b)

Figure 2. Magnetic fields distribution: a – ε =0; b – ε =0,1mm (ε is the beam deflection from the joint)

The system for joint tracking. Fig. 3 is the diagram of the system for automatic joint tracking with a flux gate as a sensor of the beam position relative to the joint of welding parts.

The differential flux gate is used as a sensor for measuring the beam deflection from the seam. The signal, proportionate to the external magnetic field, is formed in the measuring winding of the flux gate at frequency 2ω, where ω is the frequency of the flux gate excitation. At this frequency (2ω), it is important to differentiate signals for measuring constant and slowly changing magnetic fields.

At the constant electron beam current, the vertical component H_V of the magnetic field will also be constant. But, under the control principle, it is much smaller than the magnetic fields of the welded parts and welding tools as well as the magnetic filed of the Earth. Under those conditions, it may be impossible to get a signal (H_V) proportionate to the beam deflection from the joint. Additionally, significant external fields may lead to the fluxgate core saturation and make the device unusable.

If the electron beam current is included in the variable component with frequency Ω, then a component with the same frequency will be in the range of the measured magnetic fields and components with side frequencies (2ω ± Ω) (Fig. 4) will be present in the signal range of the measuring winding.

IOP Conf. Series: Materials Science and Engineering **467** (2019) 012009 doi:10.1088/1757-899X/467/1/012009

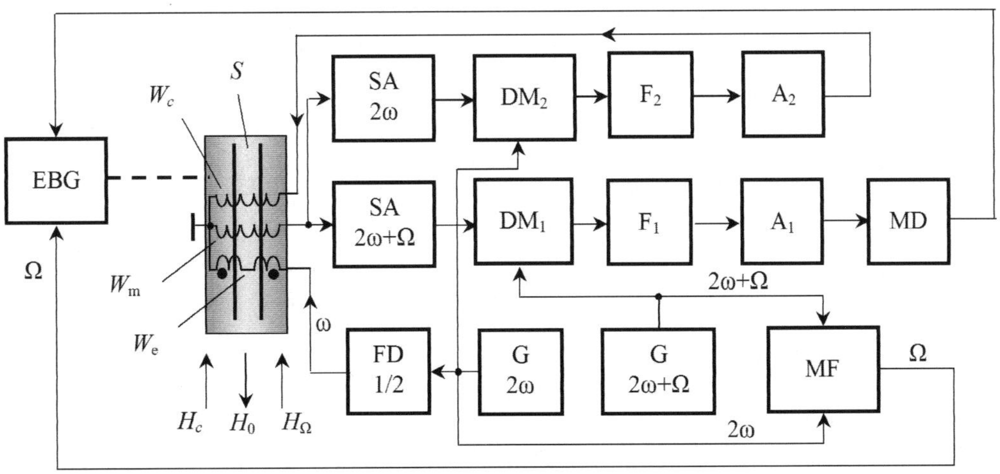

Figure 3. Structural diagram of the device for automatic beam control: S – sensor (differential flux gate); W_m - measuring winding; W_e – excitation winding; W_c – compensation winding; SA (2ω+Ω) – selective amplifier (frequency 2ω+Ω);
DM_1 – demodulator; F_1 – filter; A_1 – power amplifier; MD - motor drive; EBG – electron-beam gun; SA (2ω) - selective amplifier (frequency 2ω); DM_2 – demodulator; F_2– filter; A_2 – power amplifier; FD(1/2) - frequency divider by two; G(2ω) - frequency generator 2ω; G(2ω+ Ω) - frequency generator (2ω+ Ω); MF - frequency mixer; H_Ω - the magnetic field strength is proportionate to the deflection of the beam from the joint; H_c - magnetic field compensation; H_0 - magnetic field disturbance

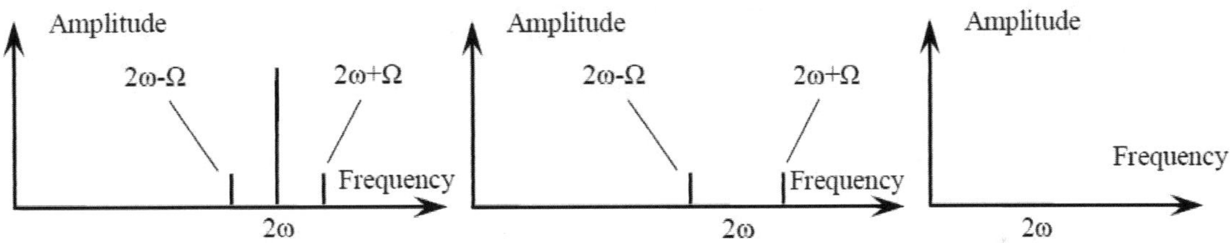

Figure 4. Spectrum components of the Fluxgate measuring winding signal: a) - $\varepsilon \neq 0$; Spectrum component of the measuring winding signal with uncompensated frequency 2ω; b) - $\varepsilon \neq 0$; Spectrum component of the measuring winding signal with compensated frequency 2ω; c) - $\varepsilon = 0$; spectrum component of the signal with frequency
2ω - compensated

To improve disturbance resistance of this system, the signal for the beam and joint misalignment is formed on the side frequency (2ω + Ω), where Ω is the electron beam current modulation frequency. For this, the output of the fluxgate measuring winding (W_m) is attached to the selective amplifier SA_1, which resonates with frequency (2ω+Ω). Then the signal is straightened with the demodulator DM_1 which has a base input attached to the generator G(2ω+Ω). After the filtration (through filter F_1), the constant current proportionate to the beam deflection from the joint goes through an amplifier (A_1) and enters the motor drive (*MD*) of the electron beam gun (*EBG*), which moves with the beam to eliminate the deflection.

The compensation channel of the constant and slowly changing magnetic fields (H_0) in the fluxgate chamber are included in the device. These fields can be caused by the residual magnetism of the welded parts and welding equipment and may lead to the saturation of the fluxgate core. When these fields appear in the measuring winding (W_m) signal range, a component emerges with the frequency 2ω and the amplitude proportionate to the amount of the current and the phase determines its direction. This component is distinguished by the selective amplifier SA_2 tuned in to resonate with 2ω frequency. Then the signal is straightened with the demodulator DM_2 which has its base input connected to the frequency generator G(2ω). After that, as the constant current passes through the filter F_2 and the amplifier A_2, the signal enters the compensating winding (W_c), where the magnetic filed (H_c) is formed, which compensates for the interference of the external fields in the fluxgate chamber.

IOP Conf. Series: Materials Science and Engineering **467** (2019) 012009 doi:10.1088/1757-899X/467/1/012009

The electron beam current modulation signal with frequency Ω is formed by the mixer (*MF*), whose inputs receive signals coming from generators with frequency 2ω and frequency (2ω+Ω). The fluxgate excitation signal is formed by the frequency divisor by 2, which has an input connected to the output of the generator with frequency 2ω. The excitation signal is delivered to the fluxgate excitation winding W_e with the frequency divisor by 2 *FD*.

The device was tested in vacuum conditions. The error of the beam alignment with the joint did not exceed 0, 3 mm, which is quite acceptable for the Electron Beam Welding in air.

Conclusions

1. We propose a method of electron beam positioning relative to the joint during the electron beam welding in the air. This method is based on identification of the magnetic fields and coordinates of the electron beam.
2. The proposed method allows controlling the electron beam location and correcting it with the help of the automated system for joint tracking with no errors.
3. The noise immunity of the proposed system is improved, as the useful signal has the frequency different from frequencies of parasite magnetic fields.
4. During electron beam welding of dissimilar materials, when the beam is at the joint, the vertical component of the magnetic field does not have a zero value due to inequality of currents in the parts. This is explained by differences in the materials conductivity. In this case, the current strength, proportional to the strength of the vertical component of the magnetic field, needs to be compensated at the input of the electron beam gun motor drive.

References

[1] Braverman V., Belozertsev V., Veisver T., Goryashin N., Lelekov A., Bashenko V. 2012 The device with a fluxgate sensor for joint tracking *Int. Conf. EBT'12.* Varna, Bulgaria, "E+E", 5-6, pp 58-63.

[2] Braverman V., Laptyonok V., Belozertsev V. Fluxgate sensor for joint tracking. Copyright certificate No. 941056 USSR, international classification index : B23K9/10, published on 07.07.82, bulletin No. 25.

[3] Braverman V., Laptyonok V., Belozertsev V. Method for measuring deviations of the electrode from the seam. Copyright certificate No. 941057 USSR, international classification index: B23K9/10, published on 07/07/82, bulletin No. 25.

IOP Conf. Series: Materials Science and Engineering **467** (2019) 012010 doi:10.1088/1757-899X/467/1/012010

Factor analysis of inelastic electron scattering cross section spectra of $FeSi_2$

A. Yu. Igumenov[1], A. S. Parshin[1], V. O. Kanzychakova[1], A. M. Demin[2], T. A. Andryushchenko[1], Yu. L. Mikhlin[3], O. P. Pchelyakov[4], V. S. Zhigalov[5]

[1]Reshetnev Siberian State University of Science and Technology, 660037, Krasnoyarsk, Russia.
[2]Emperor Alexander I St. Petersburg State Transport University, 190031, Saint Petersburg, Russia.
[3]Institute of Chemistry and Chemical Technologies of the Siberian Branch of the Russian Academy of Sciences, 660036, Krasnoyarsk, Russia.
[4]Rzhanov Institute of Semiconductor Physics of the Siberian Branch of the Russian Academy of Sciences, 630090, Novosibirsk, Russia.
[5]Kirensky Institute of Physics of the Siberian Branch of the Russian Academy of Sciences, 660036, Krasnoyarsk, Russia.

E-mail: aparshin2010@mail.ru

Abstract. Iron disilicide is widely used in creation of such nanotechnology devices as photoelectric converters. The investigation of iron silicide $FeSi_2$ with the method of inelastic electron scattering cross-section spectroscopy was carried out. The decomposition of inelastic electron scattering cross-section spectra of $FeSi_2$ to bulk and surface energy loss components using factor analysis was carried out. The amplitude of energy loss components can be used for identification of their origin.

1. Introduction

An iron disilicide system is promising for basic research and practical applications in devices of nanoelectronics, spintronics and photonics [1]. Elemental analysis of Fe-Si structures is complicated by the formation of the silicides because the chemical shifts in X-ray photoelectron spectra (XPS) practically do not depend on the iron silicide composition [1, 2] and bulk plasmon energy in reflection electron energy loss spectra (REELS) obtained by different authors give conflicting results [3].

In this paper the results of iron disilicide investigation using inelastic electron scattering cross-section spectroscopy was carried out. The decomposition of inelastic electron scattering cross-section spectra of $FeSi_2$ to bulk and surface energy loss components using factor analysis was carried out. The amplitude of energy loss components can be used for identification of their origin.

2. Experimental part

Iron silicide $FeSi_2$ was produced by melting a mixture of iron and silicon in atomic ratio of 1×2 under high vacuum conditions using the vacuum deposition installation UVN-2M-1. The mixture was maintained at the melting temperature for 15 min, after which the annealed alloy was crushed and again melted for 15 min. The 1 mm thick washers were cut of bulk samples. Before the spectroscopic investigations the washers were polished.

The spectroscopic experiments were carried out using a photoelectron spectrometer SPECS (German production) equipped with the spherical energy analyzer PHOIBOS MCD9, double anode X-ray tube as an X-ray source, and a Microfocus EK-12-M electron gun (STAIB Instruments) for excitation of the REEL spectra. The surface contaminations, protecting layers and oxide layers were removed applying Ar^+ ion etching (accelerating voltage of 2.5 kV, ion current of 15 μA) with an ion source IQE-12/38 (SPECS) in the analytical chamber of the spectrometer before the REELS measurements; the completeness of the procedure was controlled via the relevant photoelectron and Auger spectra.

REEL spectra were acquired in the interval of 150 eV below the elastic peak with an energy step of 0.1 eV. Energy loss T was calculated as the difference between the primary electron beam energy E_0 (zero loss) and the reflected electron energy E, $T = E_0 - E$. The primary electron beam energies were 300, 600, 1200, 1900, 3000 eV, and the full width at half-maximum (FWHM) of the primary electron beam was less than 1 eV.

The inelastic electron scattering cross section spectra (so-called $K\lambda$-spectra [4]) which are the products of inelastic mean free path λ and inelastic scattering cross-section $K(E_0, T)$ were calculated from the experimental reflection electron energy loss spectra with the software package QUASES^{TM} XS REELS (Quantitative Analysis of Surfaces by Electron Spectroscopy cross section determined by REELS) [5] based to the algorithm suggested in [6]. Inelastic electron scattering cross section spectra indicate the probability of electron energy loss T in a single scattering event. $K\lambda$-spectra can be used for quantitative determination of element concentrations in two-component composite structures. The ability of quantitative determination of atomic element concentrations in Fe_xSi_{1-x} [7, 8], Mn_xSi_{1-x} [7] and Ge_xSi_{1-x} [10] systems were investigated earlier with the dependence of the maximum value of $K\lambda$-spectra for the standard samples.

The results of applying FA for investigation of inelastic electron scattering cross section spectra of Si, SiO_2, GaAs are presented in [11,12]. The procedure of FA contains two main steps. The first step (principal component analysis (PCA)) yields a minimum number N of principal components (abstract factors). These components are necessary to adequately describe the data. This procedure reduces the data matrix dimensionality into the lowest dimensionalities which are sufficient to adequately describe the data. The second step (target transformation (TT)) transforms the main N abstract factors into physically meaningful factors.

In the PCA step, the Kλ-spectra are placed into a data matrix. The m measured with different E0 Kλ-spectra, having p energy data points were arranged in columns to form a data matrix D(p×m). From this, the covariance matrix $D\cdot D^T=(p\times p)$ was constructed by multiplying the data matrix D by its transpose. The eigenfunctions were determined by finding a matrix U, which satisfies $[D\cdot D^T]\cdot[U]=[x]\cdot[U]$. This is an eigenvalue problem of the covariance matrix $D\cdot D^T$ to obtain the diagonal matrix of eigenvalues $[x]$ and the matrix of eigenvectors [U]. The resulting eigenvectors were arranged in order of magnitude of the corresponding eigenvalues. The data matrix D can be expressed as a linear expansion of these principal components which are also called abstract factors. Only the abstract factors which correspond to the large eigenvalues are needed to get a good account of the data, whereas, those with smaller eigenvalues can be neglected. Although the abstract factors give a perfect mathematical model of the data matrix D, they have no physical meaning.

In the second step, we obtain physically meaningful factors by TT of the principal abstract factors. Transformation can be performed by means of target rotation. This rotation corresponds to multiply abstract factors with a transformation matrix to give new factors. In the "present" case, we have two major factors. This means the factor space has two dimensions, and a two-dimensional transformation matrix, involving an orthogonal rotation angle θ,

$$T = \begin{pmatrix} \cos(\theta) & \sin(\theta) \\ -\sin(\theta) & \cos(\theta) \end{pmatrix}$$

is first applied for a TT.

It is well known that bulk-loss component can be described by a formula of Tougaard. In Ref. [13-19] the method of inelastic electron scattering cross-section spectra fine structure analysis was suggested. The fine structure analysis is based on fitting $K\lambda$-spectra to the 3 parameters Lorentzian-type formula of Tougaard [20]:

$$\lambda K = \frac{BT}{\left(C - T^2\right)^2 + DT^2}.$$

B, C, D are fitting parameters and have special values for different elements [20]. The peak intensity depends on parameter *B*, peak energy depends on parameter *C*, peak width and, indirectly, peak energy depend on parameter *D*.

With this orthogonal rotation, we obtain the first dominant real component corresponding to the bulk-loss component. We applied these criteria for the bulk-loss component determination and changed the factor axes rotation angle until the first factor fitted well to a formula of Tougaard.

If the second component corresponding to the surface-excitation component given by this orthogonal rotation is not adequate from the physical viewpoint, an additional oblique rotation (with rotation angle φ) of one axis is done to get the second real component. After the coordinate rotation of two primitive abstract factors, the real factors with physical meaningful features are thus obtained.

3. Results and discussion

3.1. $FeSi_2$ inelastic electron scattering cross-section spectra

Fig. 1 (a) shows the inelastic electron scattering cross-section spectra for $FeSi_2$ silicide. The main peak energy increases monotonically with increasing of primary electron energy in inelastic electron scattering cross-section spectra (Fig. 1 (b)). With increasing of primary electron energy increases the depth analysis, resulting in decreasing of surface-like excitation intensity and increasing of bulk-like excitation intensity.

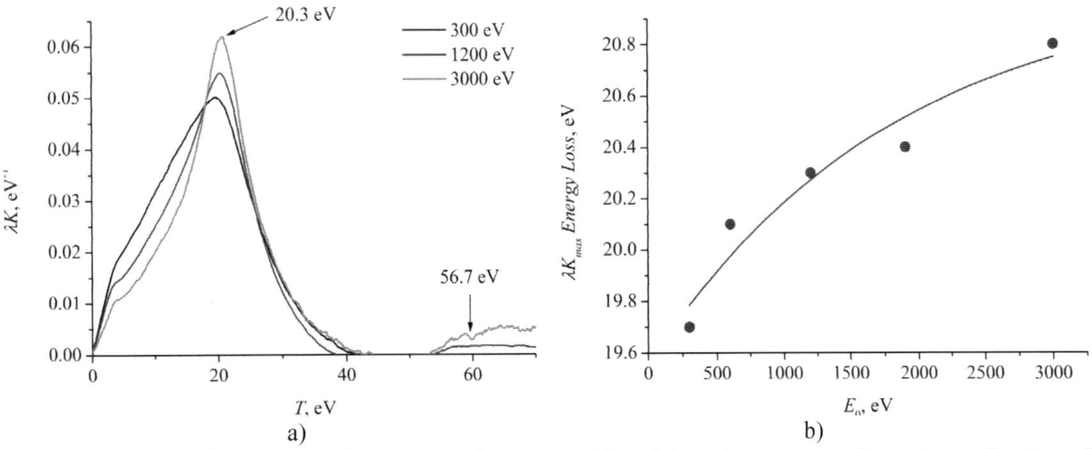

Figure 1. Inelastic electron scattering cross-section spectra (a) and the primary energy dependence of main peak loss energy (b)

This intensity redistribution causes a change of the whole spectrum shape and peak position. The presence of unresolved surface plasmon in *Kλ*-spectra of iron silicides manifests itself in decreasing of FWHM. Further, to isolate the surface and bulk contributions, the FA will be applied.

3.2. Factor analysis of $FeSi_2$ inelastic electron scattering cross-section spectra

Fig. 2 (a) shows the reconstruction of *Kλ*-spectra of $FeSi_2$ for E_0 = 300 eV (a) and E_0 = 3000 eV (b) by a linear combination of surface electronic excitation and bulk-loss components

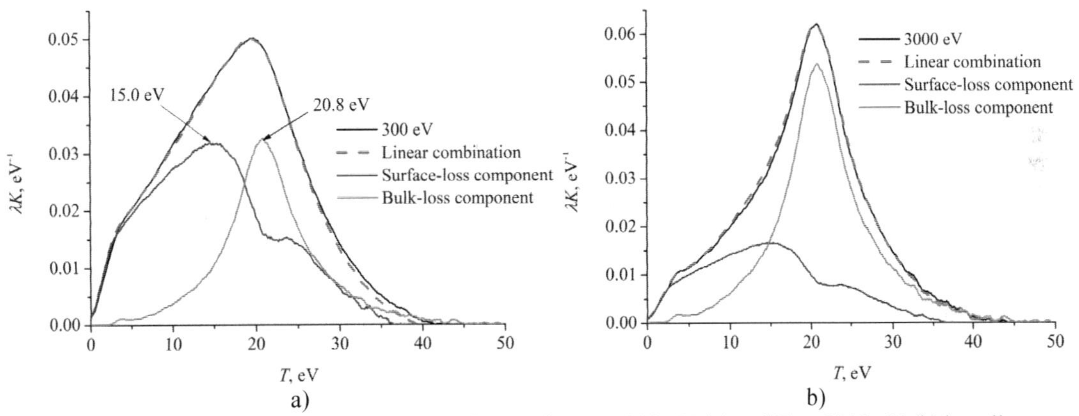

Figure 2. Reconstruction of *Kλ*-spectra of $FeSi_2$ for E_0 = 300 eV (a) and E_0 = 3000 eV (b) by a linear combination of surface electronic excitation and bulk-loss components

Obtaining real factors for $FeSi_2$, we selected the angle in a way that the bulk-loss component should be well-approximated to the Tougaard peak, similarly done for the surface one, but it turned out much worse. This is because the surface contribution is unresolved. The obtained values of the energies of the bulk and surface components are close to the literature values for the energy of the bulk and surface plasmons [3].

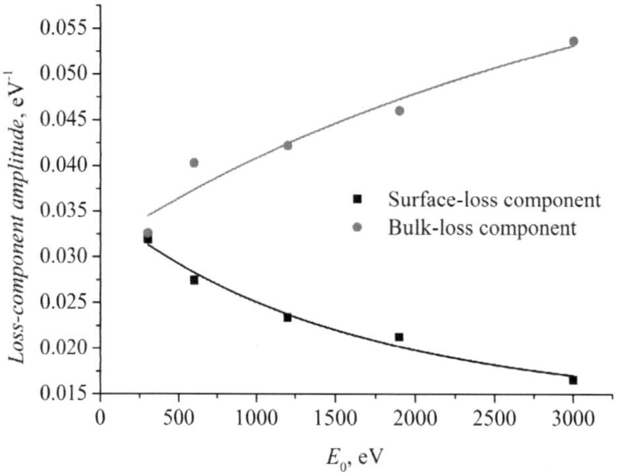

Figure 3. The primary energy dependences of the loss-component amplitude

The bulk-loss component has gradually increasing primary energy dependence of amplitude which indicate its volume origin (Fig. 3). The surface-loss component has gradually decreasing primary energy dependence of amplitude. The surface-loss component was unresolved in $K\lambda$-spectra for $FeSi_2$ but FA allows determining its energy, intensity and origin.

4. Conclusions

The inelastic electron scattering cross-section spectra for silicide $FeSi_2$ were investigated. It was shown that these spectra contains unresolved surface energy losses. Decomposition of $FeSi_2$ to surface and bulk energy loss components using factor analysis allowed to identify the unresolved energy losses, determine their energies, intensities and origin. It is shown that the amplitude of energy-loss components can be used for identification of their origin.

References

[1] Ohtsu N., Oku M., Satoh K., Wagatsuma K., 2013 *Appl. Surf. Sci.* **264** pp 219-224.
[2] Ohtsu N., Oku M., Nomura A., Sugawara T., Shishido T., Wagatsuma K. 2008 *Appl. Surf. Sci.* **254** pp 3288-3294.
[3] Lifshits V. G., Lunyakov Yu. V. 2004 Electron Energy Loss Spectra of Surface Phases on Silicon *Vladivostok: Dal'nauka.*
[4] Gergely G. *2002 Prog. Surf. Sci.* **71** pp 31-88.
[5] Tougaard S. QUASES – Software packages to characterize surface nano-structures by analysis of electron spectra http, available at: http://www.quases.com/.
[6] Tougaard S., Chorkendorff I. 1987 *Phys Rev. B* **35** pp 6570-6577.
[7] Parshin A. S., Pchelyakov O. P., Dolbak A. E., Ol'shanetskii B. Z., 2013 *Journal of Surface Investigation. X-ray, Synchrotron and Neutron Techniques* **7** pp 505-508.
[8] Parshin A. S., Aleksandrova G. A., Dolbak A. E., Pchelyakov O. P., Ol'shanetskii B. Z. , Ovchinnikov S. G. , Kushchenkov S. A. 2008 *Tech. Phys. Letters* **34** pp 381-383.
[9] Parshin A. S., Kushchenkov S. A., Aleksandrova G. A., Ovchinnikov S. G. 2011 *Tech. Phys.* **56** pp 656-661.
[10] Parshin A. S., P'yanovskaya E. P., Pchelyakov O. P., Mikhlin Yu. L. , Nikiforov A. I., Timofeev V. A. , Esin M. Yu. 2014 *Semiconductors* **48** pp 224-227.
[11] Jin H., Shinotsuka H., Yoshikawa H., Iwai H., Tanuma S., Tougaard S. *Journal of applied physics* 2010 **107** pp 083709.
[12] Jin H., Shinotsuka H., Yoshikawa H., Iwai H., Arai M., Tanuma S., Tougaard S. *Surface and Interface Analysis* 2013 **45** pp 985–992.
[13] Parshin A.S., Igumenov A.Yu., Mikhlin Yu.L., Pchelyakov O.P., Zhigalov V.S., 2016 *Physics of the Solid State* **58** pp 908-914.
[14] Parshin A.S., Igumenov A.Yu., Mikhlin Yu.L., Pchelyakov O.P., Nikiforov A.I., Timofeev V.A., 2015 *Semiconductors* **49**, **4** pp 423-427.

[15] Parshin A.S., Igumenov A.Yu., Mikhlin Yu.L., Pchelyakov O.P., Nikiforov A.I., Timofeev V.A., 2014 *Vestn. Sib. Gos. Aerokosm. Univ.* **4 (56)** pp 230-235.
[16] Parshin A.S., Igumenov A.Yu., Mikhlin Yu.L., Pchelyakov O.P., Zhigalov V.S., 2016 *IOP Conference Series: Materials Science and Engineering* **122** pp 012025, 1-7.
[17] Parshin A.S., Igumenov A.Yu., Mikhlin Yu.L., Pchelyakov O.P., Zhigalov V.S., 2016 *Technical Physic* **61 (9)** pp 1418-1422.
[18] Parshin A.S., Igumenov A.Yu., Mikhlin Yu.L., Pchelyakov O.P., Zhigalov V.S., 2016 *Izvestiya Vysshikh Uchebnykh Zavedenii, Fizika* **59, 10** pp 82-86.
[19] Parshin A.S., Igumenov A.Yu., Mikhlin Yu.L., Pchelyakov O.P., Zhigalov V.S., 2016 *IOP Conference Series: Materials Science and Engineering* **255** pp 012019, 1-7.
[20] Tougaard S. 1997 *Surf. Interface Anal.* **25** pp 137-155.

IOP Conf. Series: Materials Science and Engineering **467** (2019) 012011 doi:10.1088/1757-899X/467/1/012011

Target-specific change in the liquid permeability of wood

S G Eliseev[1], V N Ermolin[2], A V Namyatov[3]

[1]Associate Professor, Reshetnev Siberian State University of Science and Technology, 660037, Krasnoyarsk, Russia
[2]Professor, Reshetnev Siberian State University of Science and Technology, 660037, Krasnoyarsk, Russia
[3]Assistant, Reshetnev Siberian State University of Science and Technology, 660037, Krasnoyarsk, Russia

E-mail: s-555s@yandex.ru

Abstract. The article presents the findings of the study on the possibility of using the wound reaction of wood for controlled formation of the material's texture. It was determined that the wound reaction of the living wood cells is an effective mechanism for artificial decrease in the natural permeability of birch. Herewith, the process of decreasing the wood permeability can be controlled and selective due to the local removal of free moisture in the parts of the material designed for treatment.

1. Introduction

Currently, artificial materials imitating wood gain broader application. At the same time, the manufacturers of such materials almost do not have technological limitations in the choice of colors and textures when forming the range of products. However, despite this fact, a significant share of the market still belongs to natural wood products, which is due to their greater prestige in comparison with the products manufactured from artificial materials.

At the same time, the current demand for natural wood products does not comply with the supply available on the market. And the main deterrent factor of the supply growth is the limited availability of valuable species of wood with high decorative properties.

One of the ways to solve this problem is to expand the use of less valuable wood of widespread soft-leaved species that do not have natural decorative properties [12]. In the literature, there are quite a few examples of wood modification that allow it to be brightly colored and demonstrate an inexpressive texture due to full treatment with coloring solutions [2, 3, 4, 5, 6, 9, 11]. At the same time, such methods provide only a narrow range of opportunities for forming the texture based on the different intensity of coloration of the annual layer and wood ray areas. Local impregnation of wood using devices similar to those described in the work [10] is justified only to some extent. While when introducing a coloring solution it is difficult to avoid the approach or even closure of the impregnation fronts due to the redistribution of the solution. This in turn leads to a significant distortion of the designed texture and makes it impossible to form the same pattern along the entire length of the assortment.

In our opinion, a better controlled formation of the texture can be achieved by a target-specific change in the liquid permeability of wood in its individual sections. The mechanism of changes in permeability can be a wound reaction that occurs in the wood in case of the damage caused to the tree or any significant disturbance of the water-gas regime of the tree trunk. The essence of the wound reaction is the gradual death of living parenchymal cells with the formation of colored gum-like wound substances. The wound reaction is aimed at protecting the living sections of wood from drying out or fungal attack and leads to a significant decrease in the permeability of wood [1, 8]. As is known, living cells in heart and ripe woods are found only in the peripheral part of the trunk and only in sapwood along its entire section. Therefore, the formation of the wood texture with the use of wound reactions along the entire thickness of the assortment is possible only when using a sapwood such as birch.

2. Materials and methods

Uniform wound reaction along the entire length and section of the assortment is ensured by a moderate positive temperature while maintaining a high moisture content of the wood [1, 8]. However, to form the texture of the material, it is necessary to preserve local permeable areas for the subsequent introduction of the dye, which implies the need to inhibit the wound processes in the specific designed zones. In this study it was decided to use the wood moisture content out of the factors determining the duration of the wound reaction.

Special studies were carried out to determine the technological methods that would allow to targetedlyform permeable and impermeable zones in the wood. For this purpose, 9 samples with the length of 1.2 m with the diameter from 27 to 41 cm were made from 3 freshly cut round birch timber materials (Betulapubescens).

After multi-layer waterproofing of the ends and places of the damaged bark, three samples (one from each timber) were placed as a control for storage for 60 days at the temperature of 22°C.

In the three samples of the second batch the free moisture was removed locallyin the areas designed for impregnation according to the scheme depicted in Figure 1a.

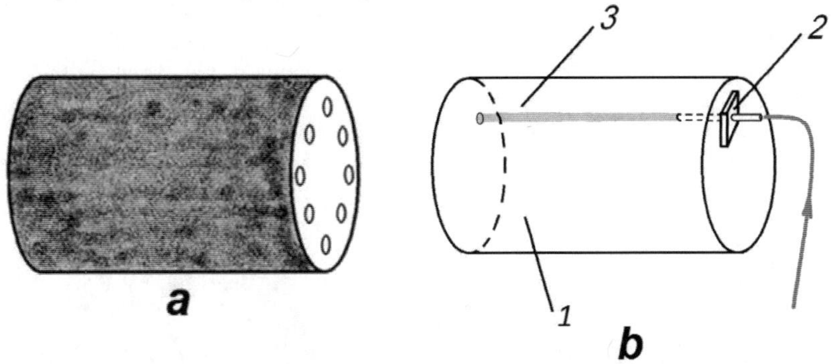

Figure 1. Preparation of the wood samples by means of local removal of the free moisture
a – the scheme of air injections; *b* – the scheme of adjustment for air injections: 1 - wood sample; 2 - air injector; 3 - frontal zone of air progression

The free moisture from cell cavities was removedlocallyusing the device shown in Figure 1b. An injector with a hole diameter of 10 mm was inserted into the end face of the sample to the depth of 20 mm, then air was fed into the wood through the injector at the pressure of 0.5 MPa. The wood was blown through with air until the exudation of water from the capillaries at the opposite end of the sample stopped.

The remaining 3 samples were subjected to partial debarking of the assortments by removing the strips of bark with the width of 50 mm and leaving unbarked 100 mm sections between them (Figure 2).

After special preparation, the samples of the second and the third batches similar to the ones of the first batch were subjected to multi-layer waterproofing of the ends and placed for storage for 60 days at the temperature of 22°C.

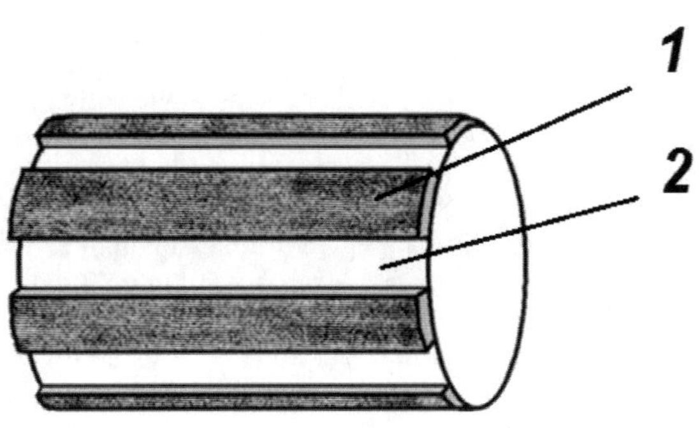

Figure 2. Scheme of the wood samples preparation by means of partial barking of the assortment
1 - unbarked area; 2 - barked area

On completion of the 60 day period of keeping the assortments at the positive temperature, the samples were trimmed by 50 mm at each end. They were further impregnated in an autoclave at the overpressure of 0.5

MPa with an aqueous solution of a textile dye. After impregnation, the character of penetration and localization of the impregnating solution was studied by splitting the samples and transversal sawing.

3. Results and discussions

The conducted experimental research has revealed the following peculiarities of the formation of impregnated and non-impregnated zones in birch samples depending on the preparation method.

The study of the wood samples the preparation of which included local removal of bound moisture showed the following. The main scope of the assortment was, as expected, impermeable to the aqueous solution of the dye, except for local areas prepared for impregnation. When considering the localized permeability areas in the cross-section, the formation of more than one permeability zone, as planned, was observed for all samples (Figure 3 (IW)). The first zone of the impregnated wood, as planned, was formed along the wood section with previously removed free moisture and showed a uniform impregnation of the dye along the entire length of the assortment. The shape of such areas was practically the same as that of the injector, and the diameters exceeded the opening of the injector by less than 3-4 mm. The preservation of the wood permeability in this case was explained by the rapid death of living wood cells in this area of wood due to their contact with the air medium without a prolonged wound reaction.

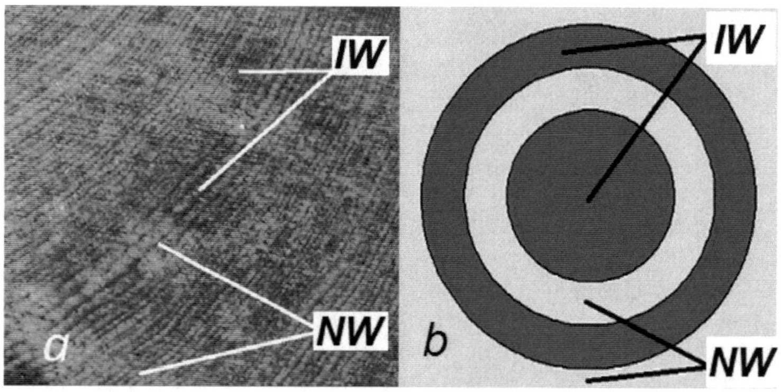

Figure 3. Formation of impregnated (IW) and non-impregnated (NW) zones in wood:
a – sample after through impregnation; *b* – scheme of forming zones with different permeability

The second permeability zone was formed as a ring with slightly uneven edges separated from the first impregnated zone by a ring of non-impregnated wood. The thickness of the impregnated wood ring reached 10 mm, and that of non-impregnated wood ring 5 mm.The rings were elliptical in shape, and in all cases the larger thickness of the rings was observed in the tangential direction (in relation to the structural directions of the wood). The formation of these rings, apparently, can be explained by the fact that when the air penetrates into certain layers of the wood, the process of wound reaction begins to violently proceed in adjacent layers, leading to the formation of the non-impregnated wood ring. In this case, the layers of wood adjacent to the wound reaction zone are the suppliers of substances necessary for the implementation of this process. And later, when the process of the wound reaction starts across the entire section of the sample, these areas are already "depleted" and do not actually participate in the process remaining permeable to liquids and gases.

The study of the samples prepared according to the scheme in Figure 2 did not reveal the designed stellular pattern in the cross section. The assortment was impregnated evenly over the entire section, except for two small sections (with the length of not more than 300 mm and a diameter of up to 40 mm) in the central part of the assortment. Presumably, the preservation of the wood permeability throughout the entire section of the sample is due to excessively rapid decrease in the moisture content of the wood and, consequently, early termination of the wound reaction.

Control samples showed insignificant penetration of the impregnating solution at slitting - not more than 20-30 mm from the ends into the depth of the assortment. The main part of the assortment though remained non-impregnated, which corresponds to the data given in the work [8].

4. Conclusions

Taking into account the findings obtained, we can make the following conclusions:

1. The wound reaction of the living wood cells is an effective mechanism for artificial decrease of the natural permeability of birch.
2. The process of reducing permeability using the wound reaction of wood can be controlled and selective due to the application of a simple and technological method of wood preparation: local removal of free moisture from the cavities of wood cells by introducing air.
3. With the appropriate adjustment and optimization of the processes, the developed method of preparation and impregnation of wood can be successfully applied for the targeted change in the decorative properties of birch wood.

The reported study was funded by Russian Foundation for Basic Research, Government of Krasnoyarsk Territory, Krasnoyarsk Region Science and Technology Support Fund to the research project №17-48-240553.

References

[1] VakinAT, Poluboyarov OI, Solovyev V A 1980Flaws in the WoodMoscow:Forestry Industry 112 p
[2] Trubnikov NA2011Analysis of the Decorative Properties of Wood and Model Study of the Changes in the Wood Texture*Forestry Engineering Journal* No**3**pp 31-37
[3] Arai Kazushge, Vota Shigeki 1991Selective Impregnation and Colored Patterns *Makudzaigakaisi. J. Jap. Wood Res. Soc.* No. **5**pp 383 - 389
[4] Belyaev E Yu et al. 1999 Wood Coloring*Chemistry of Plant Raw Materials*No. **2**pp 5-18
[5] Kushnirskaya MTs1973Wood Coloring in Furniture Manufacture / Kushnirskaya M Ts. Moscow: Forestry Industry120 p.
[6] Improvement of the Quality of Birch Finishing and Constructional Materials by Volumetric Pulse Impregnation with Water-soluble Fire Retardants and Dyes: the author's abstract of the Candidate of Technical Sciences / Lee Khunda. Tomsk: Tomsk Polytechnic University, 2010. 20 p.
[7] Eliseev S G,Ermolin V N, Namyatov A V 2014Liquid Permeability of Birch *Coniferous Trees of the Boreal Zone* Vol. **XXXII** No 5 - 6 pp 77-80
[8] Solovyev VA1983Breathing Gas Exchange of Wood. Leningrad: Leningrad State University Publishing House 300 p.
[9] Trubnikov NA 2009Development of the Method for Improving the Texture of the Soft-wooded Broadleaved Species by Selective Coloring and Pressing: author's abstract of the Candidate of Technical Sciences Voronezh: Voronezh State University of Forestry and Technologies 16 p
[10] Devices for Impregnation of Cut Wood: Patent 865028/29-14 USSR: MPK V 27k, 64YuZ, Kl. 38h, 4 / Chernenko SA – No. 829507/29-14; application submitted April 06, 1964; published July 30, 1964, Bul., No. 14 – 6 p.
[11] Bernat A I,ChernenkoSA 1968Wood Modification by Dyes with the Method of Ends' Impregnation under Pressure. Moscow: Forestry Industry 38 p.
[12]Shamaev VA 2010Modern State and Ways of Development of Modified Wood *Materials of the International Scientific and Practical Conference "Modern Challenges of the Mechanical Technology of Wood". Saint Petersburg*pp 11-17

IOP Conf. Series: Materials Science and Engineering **467** (2019) 012012 doi:10.1088/1757-899X/467/1/012012

Experimental research on electron-beam welding technology with a scanning electron beam

Yu N Seregin, V D Laptenok, A V Murygin, A N Bocharov

Reshetnev Siberian State University of Science and Technology, 660037, Krasnoyarsk, Russia

E-mail: avm514@mail.ru

Abstract. In this paper the question of achievement of quality welded joints in electron-beam welding is considered. The welded seam is largely characterized by its depth, width and their relation. The parallelism of the sides of the penetration zone plays an essential role, as well as the radius of the root part and other geometric parameters that affect the nature and magnitude of the welding deformations and stresses in the welded products. Traditional welding regimes entail root defects, porosity, undercuttings and fissures, and these are unacceptable in the manufacture of critical products. To obtain the required seam geometry various technological methods are used, for example, various forms of electron beam scanning with the programming of energy along the trajectory; the use of welding with different focus positions; oscillation of the electron beam; programming of change of current and focus at the beginning and end of the welding process. Nine difference trajectory scans were programmed in order to manage the energy distribution of the electron beam across the heating spot.

The effect of various forms of energy distribution across the heating spot on the process of formation of a welded seam and its characteristics was studied in this research. The quality of welded joints was evaluated by the results of the x-ray control of welded joints and microsections of the cross-section and longitudinal section of the welded joints.

1. Introduction

The quality of a seam in the process of electron-beam welding is determined by a combination of the technological and energy parameters of the process. Maintaining the energy parameters of the welding process at the required level ensures the constancy of the geometric dimensions, as well as of parameters relating to structure and strength among others under the same technological conditions. However, the possibility of forming a unique close-range form with a minimal metal content of the bath comes into conflict with the achievement of stable operating parameters of the welded joint. Violation of the optimal mode of electron-beam welding often leads to the appearance of defects in the seams, even on well-welded materials. These violations are found in all methods of fusion welding and non-welds, undercuts, sagging and increased spattering are all widely known. However, other specific defects also exist; there are root defects, extended cavities at the root of the seam, "median" cracks and seam deflections from the joint due to residual or induced magnetic fields [1]. The root defects of the seam are inherent in all methods of welding with highly concentrated energy sources. The root defects have a hydrodynamic nature of formation and they are connected with the peculiarities of the metal transfer in a weld pool.

To prevent root defects it is necessary to form a steam-dynamic channel with a sufficiently wide lower part and a rounded bottom. The form of the channel is changed by the form of power distribution of the electron beam along the heating spot [3-5]. A selected focus level of the beam [6, 7] is not able to completely suppress the root defects, and in any event the probability of incomplete fusion remains quite high. In addition, the triangular form of the deposit leads to considerable deformations and stress. In practice, welding with full penetration is used to eliminate root defects [8-10] but this method is not applicable for closed seams. The most efficient way to control the distribution of energy affecting the formation of the penetration channel is by scanning the electron beam. The following beam scans are widely used: longitudinal, transverse, X-shaped, circumferential, ellipse and arc [11]. The calculation methods for selecting the beam sweep shape have not yet been developed. At the same time, a rather large amount of experimental data is accumulated. However, the practical recommendations do not always coincide. The latter facts are largely determined by the specifics of the equipment used for scanning of the beam with limited capabilities and imperfections in the methods of investigation. The relationship between the scanning parameters and the beam focusing level, the geometry of the initial stationary

IOP Conf. Series: Materials Science and Engineering **467** (2019) 012012 doi:10.1088/1757-899X/467/1/012012

electron beam, the welding speed, and the level and the pulsation frequency of the beam parameters are not taken into account.

The effect of the sweep is shown in the change in both the instantaneous distribution of the power density of the electron beam and this distribution averaged over a period. Accordingly, the nature of the hydrodynamic processes and the configuration of the weld pool change. When welding thick metal with a horizontal beam, it is possible to expand the diameter sharply and increase the stability of the channel in the weld pool, which has a favourable effect on the stability of the formation of seams, i.e. the spattering of molten metal decreases, and the melt is prevented from flowing out of the bath. As a result of changing the seam form, the likelihood of cracking, root defects and extended cavities decreases.

At relatively large scanning frequencies (1000-2000 Hz) the beam power is distributed along the scanning path uniformly without leading to fluctuations in the depth of penetration. Therefore, in practice such sweep frequencies are commonly used.

In the Siberian Research Centre for Electron Beam Technologies (SICELT) of Reshetnev Siberian State University of Science and Technology, an industrial sample of equipment for electron-beam welding has been developed and created with a converter source of accelerating voltage [12]. In this equipment new technological capabilities have been realized to control the energy distribution of the electron beam over the heating spot. In this equipment a microprocessor block for the functional scanning of the electron beam is used allowing the beam to be scanned along various programmable trajectories. In the functional scanning block high-speed channels of beam deflection are applied to the coordinates X, Y with a bandwidth of frequencies up to 60 kHz. The electron-beam equipment CEP-2 has been modernized in the replacement of deflecting and focusing systems by high-speed ones.

As a result, a periodically collapsing steam and gas channel, typical for traditional electron-beam welding technologies, is, with the appropriate scanning paths, transformed into a stable vapour-gas cavity, extending to a significant part of the penetration depth. This action leads to a change in the conditions for the formation of the welded seam, an increase in the efficiency of the welding process, a change in the shape of the welded joint, and an improvement in the quality of the welded joints [13].

2. Material and methods

Preliminary investigations based on the use of these technological capabilities have revealed the promise of new scanning paths in order to improve the technology of electron-beam welding and improve the quality of welded joints.

Experimental investigations were performed on an ELU-5 electron-beam unit with electron-beam equipment and a CEP-2 gun with a Ø 4.7 mm tablet. Welding was performed on annular specimens of Ø 300 mm with a thickness of 27 mm made of AMg-6-HH material. The welding speed was 30 m / hour.

The purpose of this experimental research was to study the influence of various forms of energy distribution on the heating spot on the process of formation of the welded seam and its characteristics.

The quality of welded joints was evaluated by the results of x-ray inspection of welded joints and microsections of the cross-section and longitudinal section of the welded joints.

For managing the distribution of the electron beam across the heating spot we programmed nine different scanning trajectories of electron beam scanning paths. The scan numbers correspond to the following images:

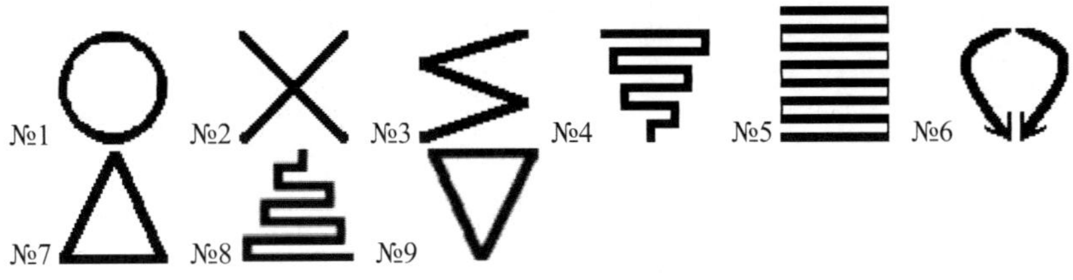

It was possible to change and control the scanning amplitudes by "x" and "y", by the scanning frequency and by defocusing the beam through the dynamic focusing channel.

Welding was executed at an accelerating voltage U = 25 kV, and the currents of the electron beam were $I_{л} = 100, 150, 200, 250\ mA$.

IOP Conf. Series: Materials Science and Engineering **467** (2019) 012012 doi:10.1088/1757-899X/467/1/012012

To exclude the effect of heating of the sample on the formation of the welded joint, the following welding was performed on a cooled model. More than 200 models were welded.

Regime 1 - without scanning.

№	1	2	3	4
The current defocus, mA	-16	30	30	20
The beam current, mA	180	150	180	200

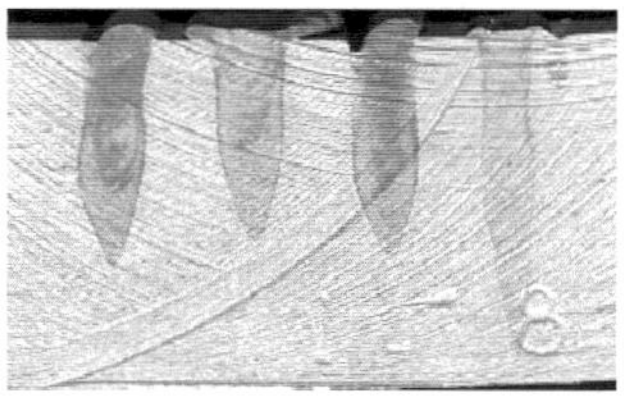

When scanning is not used, the welding was performed with different values for the beam current and with different defocusing currents along the dynamic focusing channel. As can be seen from the microsections, the form of the welded joint has a type of close-range penetration with a sharp root of the seam, with root defects and instable penetration.

Regime 2 – scanning № 1.

№	1	2	3	4	5
The amplitude X, mm	0,8	1,0	1,2	1,2	2,5
The amplitude Y, mm	1,5	2.0	2,2	2,2	2,5
The beam current, mA	150	150	200	250	250

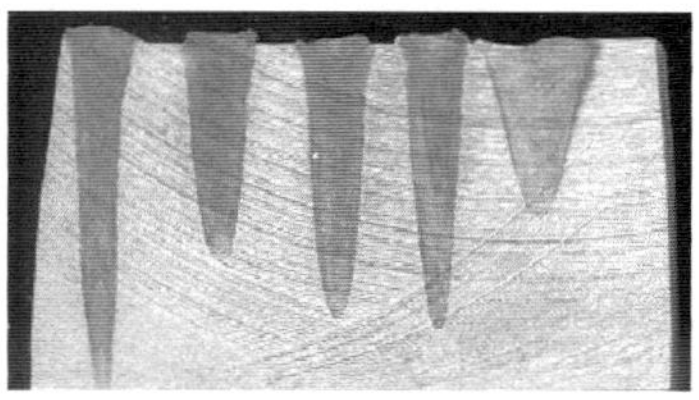

The form of the seam is cosinusoidal with a radius of curvature at the root of the seam of 0.1-0.3 mm. In addition, there are root defects and significant instability in the penetration depth. This mode is also unsuitable for practical use.

Regime 3 – scanning № 8.

№	1	2	3	4	5	6	7
The amplitude X, mm	1,4	1,6	1,4	1,6	0,8	1,0	0,6
The amplitude Y, mm	1,4	2,5	2,5	2,5	2,5	2,5	1,2
The beam current, mA	250	100	150	150	200	200	200

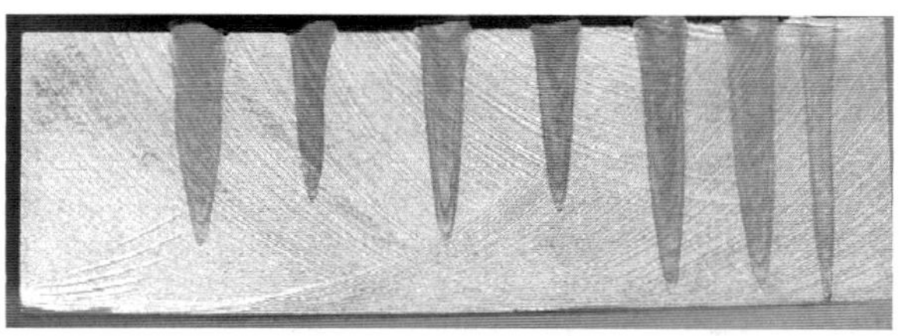

The form of the seam is approaching a cylindrical form in its upper part, which is 2/3 of the penetration depth. The lower part of the seam is close to being conical with a rounding radius of 0,3-1 mm depending on the scanning amplitudes. This mode is of interest for further research and optimization of the welding process.

Regime 4 – scanning № 5.

№	1	2	3	4	5	6
The amplitude X, mm	0,6	0,8	1,0	1,2	1,2	1,6
The amplitude Y, mm	1,2	1,6	2,0	2,0	2,0	2,5
The beam current, mA	200	150	100	150	200	200

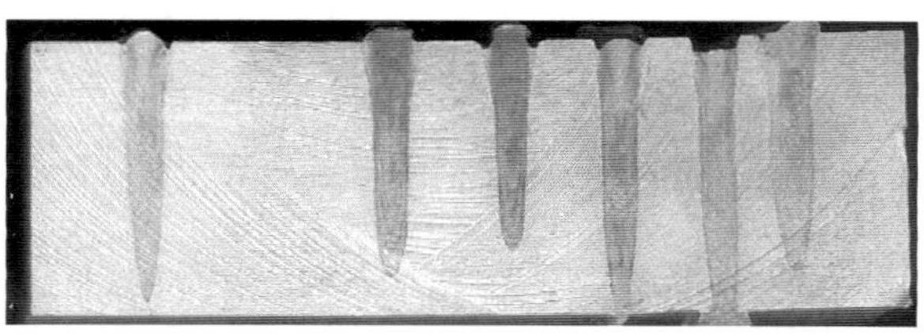

The form of the seam is from conical to nearly cylindrical with a radius of fillets from 0.3 to 1 mm. depending on the

IOP Conf. Series: Materials Science and Engineering **467** (2019) 012012 doi:10.1088/1757-899X/467/1/012012

scanning amplitude. It is possible to obtain welds without root defects. This regime is of considerable interest for further investigation.

Regime 5 – forward plane scanning №4

№	1	2	3	4	5	6
The amplitude X, mm	0,6	0,8	\1,0	1,2	1,4	1,4
The amplitude Y, mm	1,2	1,6	2,0	2,4	2,8	3,2
The beam current, mA	150	150	150	150	220	200

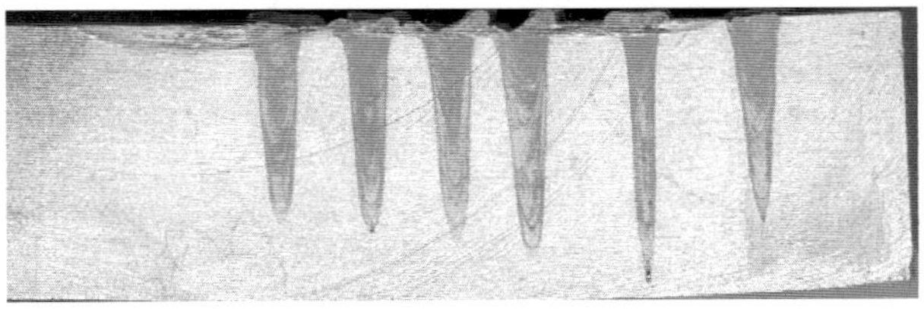

The form of the seam is mostly conical. The asymmetry of the seam relative to a vertical line with a pronounced slope is observed. There are root defects. This scanning mode is not of practical interest in terms of its use.

Regime 6 – scanning №2.

№	1	2	3	4	5	6
The amplitude X, mm	0,6	0,8	1,0	1,2	1,2	1,2
The amplitude Y, mm	1,2	1,6	2,0	2,4	2,8	3,2
The beam current, mA	200	200	200	200	250	200

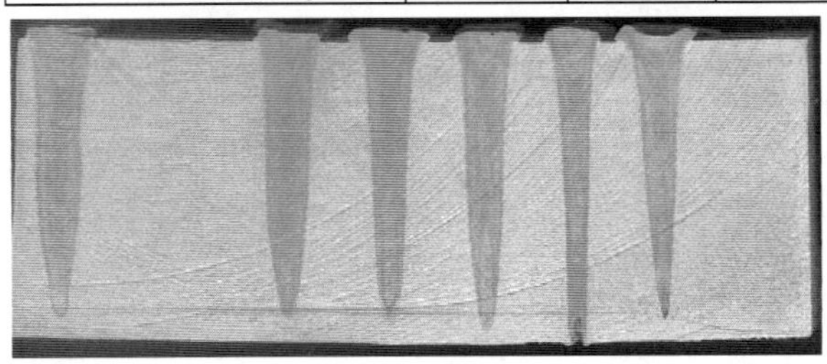

The form of the seam is nearly triangular with an acute angle in its root. There are root defects and an instable penetration depth. This mode is not of interest for further analysis or optimization.

Regime 7 – scanning №4 angle forward.

№	1	2	3	4	5	6	7	8	9
The amplitude X, mm	0,6	0,8	1,0	1,0	1,2	1,2	1,2	1,2	1,2
The amplitude Y, mm	1,2	1,6	2,0	2,0	2,4	2,8	3,2	3,2	3,6
The beam current, mA	150	200	250	200	250	250	250	200	250

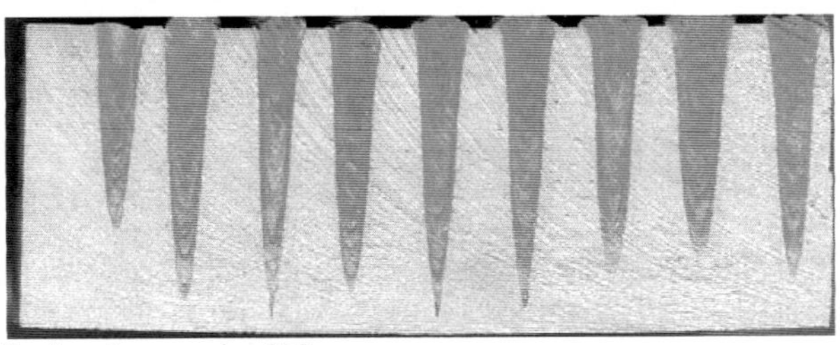

The form of the seam is clearly triangular with a sharp apex at the root. There are root defects and instable penetration. This regime is of no interest for further research.

Regime 8 – scanning №3.

№	1	2	3	4	5	6	7	8
The amplitude X, mm	1,6	1,2	1,2	1,2	1,6	1,0	0,8	0,6
The amplitude Y, mm	3,2	3,2	2,8	2,4	2,0	2,0	1,6	1,2
The beam current, mA	250	250	250	250	200	200	200	150

IOP Conf. Series: Materials Science and Engineering **467** (2019) 012012 doi:10.1088/1757-899X/467/1/012012

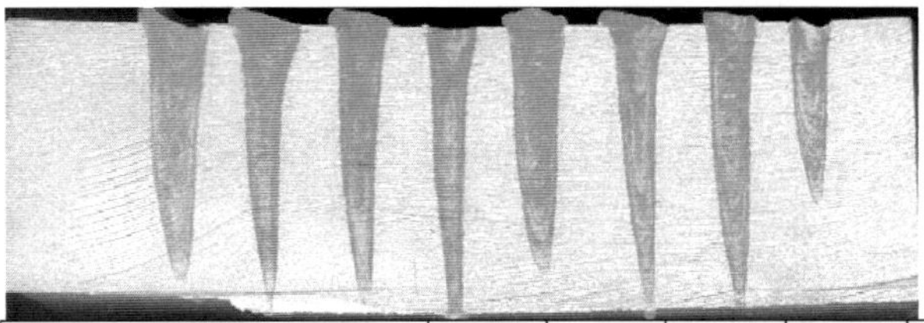

The form of the seam is irregular. There are root defects and discontinuities. This mode is of no interest in practice.

Regime 9 – scanning №6.

№	1	2	3	4	5	6	7	8	9
The amplitude X, mm	0,8	1,4	2,0	0,8	1,6	0,6	0,8	0,8	0,8
The amplitude Y, mm	2,0	2,4	2,8	3,2	3,2	1,4	3,2	2,0	3,2
The current defocus, mA	0	0	-20	0	0	0	-15	5	-5
The beam current, mA	150	200	250	200	250	250	250	200	250

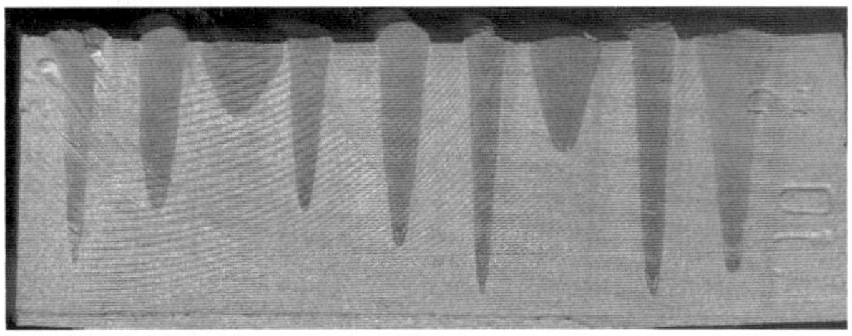

The form of the seam is triangularly peaked asymmetric. There are root defects. The regime is of no practical interest.

Regime 10 – scanning №7.

№	1	2	3	4	5	6	7
The amplitude X, mm	2,0	1,6	0,8	0,8	0,8	0,8	0,8
The amplitude Y, mm	2,0	2,4	2,8	3,2	3,2	3,2	3,2
The current defocus, mA	0	0	0	0	-5	-10	-15
The beam current, mA	150	200	200	200	150	250	250

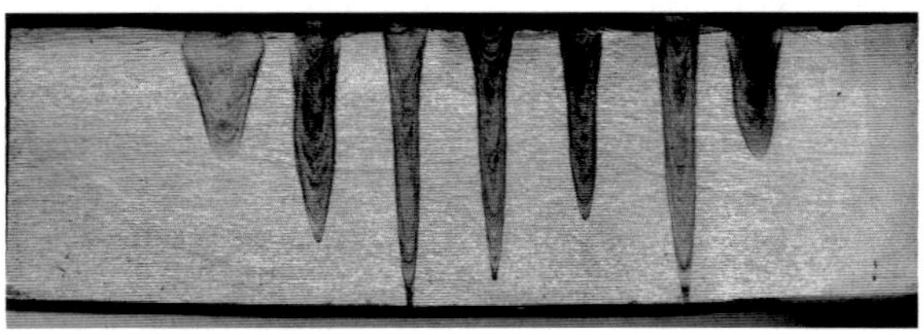

The form of the seam varies from triangular peaked to triangular blunted. The weld walls are almost parallel with through penetration. However, when the beam is taken out at the end of welding, there exists a possibility of root defects appearing.

Regime 11 – scanning №9.

№	1	2	3
The amplitude X, mm	1,6	1,6	0,8
The amplitude Y, mm	3,2	3,2	3,2
The beam current, mA	250	250	300
$v_{cв}\left[{m}/{hour}\right]$	60	60	60

1)  2)  3) 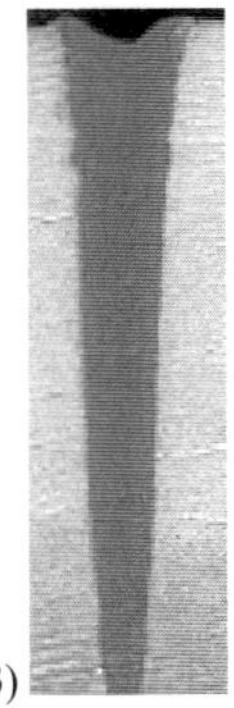

The form of the seam is triangular with a rounding at the root of a seam. It can possibly be used with through penetration. Further research is required.

3. Discussion of results

The results of the first phase of the experimental research confirmed the prospects of new technological methods for improving the quality of welded joints.

The most promising are the electron beam scans according to the forms (5), (8) and (9). In certain modes of the technological process, a more optimal form of the welded seam, a lack of root defects and a decrease in porosity are observed.

To achieve optimal technological processes it is suggested to continue the experiments with the electron beams denoted by these three scans, while varying the welding speed. The investigation of the forms of beam scanning was executed at welding speeds $v_{св} = 100 \ldots 60 \ldots 30$ m/hour.

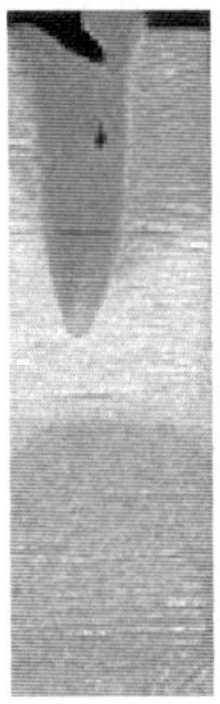

Figure 1.

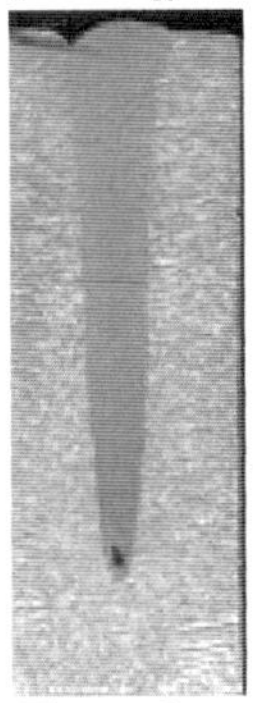

Figure 2.

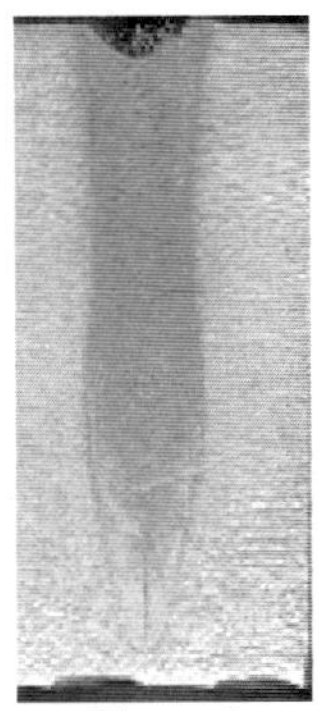

Figure 3.

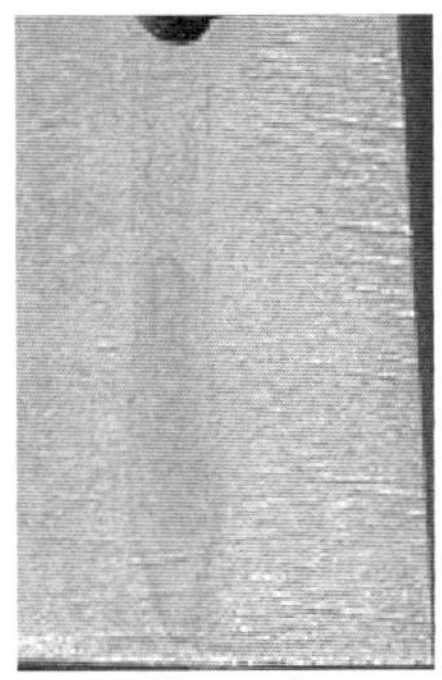

Figure 4.

Figure 1 shows the microsections of the welded joints obtained during scanning № 9 $x = 1{,}6mm;\ y = 5mm;\ v_{weld} = 100 {m}/{hour};\ I_{beam} = 0 \ldots 300mA$. The form of the seam is similar to the one received earlier (mode 11) for the same scan. The form of a seam approaches a triangular form with a radius of 0.5-1mm at the root of the joint.

Figure 2 shows the microsections of the welded joints obtained during scanning № 8 $x = 1{,}6mm;\ y = 5mm;\ v_{weld} = 100 {m}/{hour};\ I_{beam} = 200 \ldots 250mA$. In the microsection the instability of the formation of the seam at a specified speed is clearly visible.

Figure 3 shows the microsections of the welded joints obtained during scanning № 8

$x = 1{,}6mm;\ y = 5mm;\ v_{weld} = 60 {m}/{hour};\ I_{beam} = 0 \ldots 290mA$

Figure 4 shows the microsections of the welded joints obtained during scanning № 8 $x = 1{,}6mm;\ y = 5mm;\ v_{weld} = 45 {m}/{hour};\ I_{beam} = 150 \ldots 270mA$.

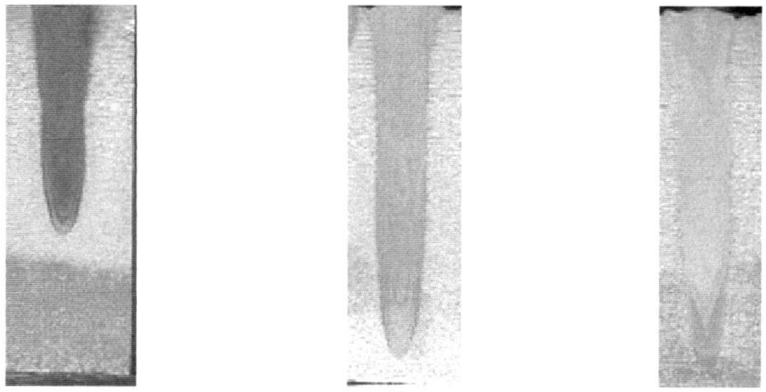

а б в

Figure 5. The microsections of the seam obtained in the following regimes:

$а)\ x = 2{,}0mm;\ y = 10mm;\ v_{weld} = 30\,{}^{m}\!/_{hour};\ I_{beam} = 178mA$

$б)\ x = 2mm;\ y = 10mm;\ v_{weld} = 30\,{}^{m}\!/_{hour};\ I_{beam} = 228mA$

$в)\ x = 2{,}0mm;\ y = 10mm;\ v_{weld} = 30\,{}^{m}\!/_{hour};\ I_{beam} = 245mA$

Figure 5 shows microsections of the welded joints obtained by scanning No. 5 at a welding speed of 30 m / h.

Analysis of the conducted investigations showed that the best results were obtained by scanning No. 5. This form of scanning makes it possible to obtain a form of the welded joints close to rectangular with almost parallel walls, while the number of stresses and deformations in the welded joints decrease. The rounding radius is 1-2 mm. In the welded seams there are no root defects at all; porosity is reduced, which is confirmed by the results of the x-ray inspection of the welded joints. Longitudinal sections of welded seams showed that the processes of formation of the welded joint proceed more stably. In comparison with traditional technology, the instability of the penetration depth is decreased by 2-3 times, shown in the form of vibrations of penetration in the root of the seam.

The output of the electron beam at the completion of the welding process can be realised only in the case of a linear decrease in the welding current without changing the parameters of the beam focusing and the scanning amplitudes. Root defects do not occur.

Experiments were conducted to stop the EBW process abruptly as a result of the switching off the accelerating voltage. In addition, the experiments have shown that at amplitudes of scanning along a joint of more than 5 mm, the vapour-gas channel is transformed into a stable cavity along the entire depth of the weld, while the thermal efficiency is increased by reducing the energy expended on the evaporation of the material.

One defect that is difficult to control in EBW with a thickness of more than 10 mm is the root defect that forms at the root of the seam as voids of different forms. With x-ray control this defect manifests itself as a chain of pores of various forms with superposition and contact. This phenomenon is not connected to the habitual pore formation and depends only on the energy distribution in the heating spot. Poor weld formation with a distorted heat spot, as a rule, leads to the formation of a root defect and the more concentrated the heating spot is, the less weldable thickness is that is affected by this defect. A welding thicknesses of over 12 mm is not guaranteed against varying intensities of damage due to this defect. In order to eliminate this defect it is necessary to control the energy distribution law in the beam heating spot and form the seam in such a way that the root of the seam will be blunted with a radius of at least 1 to 2 mm., with the voids formed by the vapour-gas channel being filled with molten metal.

4. Conclusion

The following results were achieved in the research:

1. Programming the distribution of the energy of the electron beam in the weld pool by means of corresponding sweeps of the beam allows the formation of the welded seam in electron-beam welding to be controlled.

2. An asymmetric sweep relative to the vector of the welding speed leads to the asymmetry of the seam along the depth of penetration.

3. By choosing the sweep of the electron beam, the quality of the welded joints is improved by eliminating root defects, thus reducing the porosity and instability of melting and reducing stresses and deformations in welded joints.

The reported study was funded by Russian Foundation for Basic Research, Government of Krasnoyarsk Territory, Krasnoyarsk Region Science and Technology Support Fund to the research project № 17-48-240098

References

[1] Terentyev E V, Sliva A P 2013 The choice of the minimum permissible welding speed for ELS with through penetration *Bulletin of the Perm National Research Polytechnic University. Mechanical engineering, materials science* T **15**, No. 4 pp 40-47
[2] Nazarenko O K, Kaidalov A A and others 1987 Electron-beam welding [Text] Ed. B.E. Paton. Kiev Science. dumka 256 p
[3] Belenky V Ya, Trushnikov D N, Mladenov G M, Ol'shanskaya T V 2012 Features of obtaining high-quality welded seams in electron-beam welding of high-strength steels of high thickness *Automatic welding* No. **2 (706)** pp 47-50
[4] Ol'shanskaya T V, Trushnikov D N, Belen'kii V Y, Mladenov G M 2013 Effect of electron beam oscillations on the formation of the structure and properties of the welded joint / *Welding Inrenational* Vol. **27**, iss. 11 pp 881–885
[5] Olshanskaya T V, Salomatova E S 2016 A review of modern methods of electron beam control in electron beam welding *Bulletin of the Perm National Research Polytechnic University. Mechanical engineering, materials science* **18** No 4 pp 169-186
[6] Yazovskikh V M, Trushnikov D N, Belenky V Ya 2004 The mechanism of secondary emission processes in electron-beam welding with electron beam modulation *Welding* **4** pp 21-27
[7] Method of electron beam welding: pat. 2532626 Ros. Federation: IPC B23K15 / 02 / Trushnikov DN, Belenky V.Ya., Lyalin AN, Piskunov AL, Shchavlev VE; applicant and patent holder Perm. nat. Issled. polytechnical. un-t. - No. 2013113445/02; claimed. 26.03.13; publ. 10.11.14, Bul. № 31. 10 p.
[8] Goncharov A L, Sliva A P, Portnov M A, Terentyev E V, Gribkov M.S 2013 Investigation of welded joints of dissimilar steels of magnetic circuit parts performed by ELS *Bulletin of the Perm National Research Polytechnic University. Mechanical engineering, materials science* T **15** No. 4 pp 74-87. Oscillation of the electron beam as a means of improving the formation of the root of the seam 157
[9] Boldarev E A, Pashkin A N, Bogdanov V V 2014 Production of drill bits with the use of ELS *Actual problems of aviation and cosmonautics* T. **1** No. 10 pp 91-92
[10] Medvedev A M, Semenov A M, Semenov Yu I, Sizov M M, Starostenko A A, Tsyganov A S 2016 The process of manufacturing vacuum chambers for the FAIR project using electron beam welding *Bulletin of the Perm National Research Polytechnic University. Mechanical engineering, materials science* T. **18** No. 3 pp 135-145
[11] Shilov G A, Akopyants K S, Kasatkin O G 1983 Influence of frequency and diameter of circular scanning of an electron beam on metal penetration at ELS *Automatic welding* No. **8** pp 25 - 28
[12] Seregin Yu N, Laptenok V D 2008 Automated electron beam equipment AELA 30/10. Technologies and equipment of electron-beam welding *Materials of the First St.-Petersburg International Scientific and Technical Conference. St. Petersburg. "Agency Vint-Print."* pp 137-140
[13] Terentyev E V, Sliva A P 2013 The choice of the minimum permissible welding speed for ELS with through penetration *Bulletin of the Perm National Research Polytechnic University. Mechanical engineering, materials science* T **15** No 4 pp 40-47

Method for determination of technological mode parameters of electron-beam welding based on the application of optimality criterion with the view on the weld pool uniform heating

Yu. N. Seregin, V. D. Laptenok, A. V. Murygin, V. S. Tynchenko

Reshetnev Siberian State University of Science and Technology, 660037, Krasnoyarsk, Russia

E-mail: vadimond@mail.ru

Abstract. The paper presents the results of mathematical modelling of temperature fields when the AMG-6 alloy is heated with energy sources equivalent to an electron beam in electron-beam welding. The evaluation of the results was performed using the criterion of optimality proposed by the authors. Used as a functional for calculations was the normalized standard deviation of the temperature from heating by a complex heat source consisting of mobile instantaneous point and linear energy sources. Proposed in this work are both an algorithm for finding solutions using an optimality criterion and the results of calculations for plates with thicknesses of 0.12cm and 1.2cm from the AMG-6 material, which were compared with experiments carried out under laboratory conditions.

1. Introduction

Welding is one of the most important technologies in industrial production. Depending on the production requirements, various methods of welding have been developed and are now being widely used. Among these, electron-beam welding (EBW) provides a solution to most of the problems regarding the design and manufacturing of structural metallic materials [1-5].

In the heart of electron-beam welding lies the ability of an electron beam to create on a small area of a surface high heat flux densities sufficient for heating, melting or vaporizing practically any material. This is due to the thermal effect of the radiation absorption by opaque solid materials.

The main advantages of electron-beam welding are provided by the achievement of a high energy density in the electron beam, the controllability of its parameters and the possibility of fully automating the welding process.

Reasons for the wide application of electron-beam welding include high welding speed, ability to create deep and narrow welds and a minimal value for the thermal influence zone.

2. Physical modelling of processes in electron-beam welding

Despite the fact that electron-beam welding has been used in industrial production for a long time, the physical processes in the interaction of an electron beam with metal have not been fully studied. There are several factors which have led to the creation of simplified physical models of the electron-beam welding process. These include the complex nature of interaction between the intense energy flow and the material being welded in the interaction zone in solid, liquid and gaseous states; insufficiently studied mechanism of electron beam penetration into the metal; as well as the complex dynamics of the liquid bath in the metal and the vapour-plasma flows. In most cases, instead of an accurate description in existing models, a crude approximation of the processes is used. The actual distribution of the electron beam power in the zones where it interacts with metal is the function of coordinates and time. The reason for this is the complex nature of the processes occurring during electron-beam welding.

The sufficient description of all the processes associated with welding is further complicated by the fact that there is no formulation of the processes as completely adequate general equations or statement of the corresponding initial and boundary conditions. As a result, the physical models proposed in the literature are greatly simplified and are based on the assumption of the quasi-stationary existence of the penetration channel and the weld pool.

Analysis of the processes of energy distribution injected by the electron beam into metal during welding is the basis of most attempts to relate the parameters of electron-beam welding to the geometric characteristics of welded seams. However, since the dimensions of the weld pool are small, the heat field in the rest of the welded part can be calculated as an approximation of heat propagation in solid metal heated by a point source or a linear source of heat, which gives fairly accurate results for practical use.

Based on the general theory of Rosenthal-Rykalin, in which an infinite body is heated by a moving point or linear heat source, taking into account characteristics of the electron beam, formulas and nomograms were obtained to calculate the geometrical parameters of welded seams for the electron-beam welding of thin plates [6], as well as for the depth of the liquid metal layer with an electron-beam thermal modification of the surface [7]. In this case, a high level of correlation between the calculated data and the experimental results was obtained.

The necessity of increasing the complexity of physical models and determining characteristics of an electron beam, as well as that of controlling real heat transfer in the welded part, requires creation of adequate statistical models for electron-beam welding. Statistical approach can provide a choice of process parameters for a particular welding mode and the material of the unit to be welded. A model approach to improve the accuracy of calculations applied to the conditions of mass production is given in [8].

The main problem of electron-beam welding is to obtain a reproducible cross-section of the welded seam. At present, in most cases this is achieved by the empirical selection of welding modes on control samples with cutting, preparation of sections and analysis of the weld cross section. After selecting the necessary modes for the same setting of the electron beam welding installation and on the same material, the required welds are performed. In this case, reproducibility is provided only at a specific welding installation and until the scheduled maintenance work (for example, changing the cathode).

At present, various models of thermal sources are used to calculate the dimensions of the weld pool and the heat-affected zone, which are still used to a greater extent for research purposes. Existing models include a large number of difficulties to determine parameters that depend on the welding mode. Determination of these dependencies is possible by solving the optimization problem using the appropriate experimental data. Such calculations require a lot of time and highly-qualified researchers.

The publications of Russian and foreign researchers reflect models of volumetric energy sources, which propose to predict the shape of the melting channel. The proposed models are presented both in the form of regular geometric shapes [9-16], and in more complex ones that take into account the convection in the weld pool and a functional-analytical technique for calculating temperature fields [17-19]. In [20], the model was built on the basis of collection and statistical evaluation of the experimental data for the EBW process. The task was to minimize the area of the weld. This problem was solved using a binary genetic algorithm with penalties. The solution turned out to be close enough to the global optimum. In [21] the heating source was estimated with the Levenberg-Marquardt algorithm. After this, a complete simulation of the heating process was presented. The Levenberg-Marquardt algorithm was used to analyse the sensitivity of measurements. The sensitivity analysis showed that measuring the temperature at a distance of 2.5 mm from the heating source line does not make it possible to identify the parameters of the electron beam. The study showed the expediency of using the Levenberg-Marquardt method to identify the source of heat.

It should be noted that the disadvantages of existing thermal models require current design approaches and computer modelling to be improved, and also experimental techniques for monitoring and controlling the electron-beam welding process to be developed.

When creating new products and structures from new materials using electron-beam welding, it is important to choose such parameters of the technological process that ensure the required characteristics of the welded seam. In cases of welding products with small penetration thicknesses, not so many parameters need to be selected to ensure the required characteristics of the welded seam. Such parameters include welding amperage, focal length and welding speed. Often such welding processes last for seconds and determining the required mode on analogue samples under production conditions is a rather laborious operation. Currently, the choice of these parameters is based on available modes, or on the recommendation of literature sources [9].

3. The method electron-beam welding modes finding

The authors propose a method for finding electron-beam welding modes based on the classical representation of an electron beam in the form of moving point and linear instantaneous energy sources. Together with the models of the thermal process accepted for consideration, their limitations are taken into account and the temperature limitation is introduced, at which the appropriate thermal coefficients are used. The essence of the proposed method is application of the optimality criterion for calculating the basic parameters of the electron-beam welding technological mode. As a functional in the criterion, the mean-square deviation function [22] of the normalized temperature of metal heating T_{Hi} at the final moment of time $t = t_1$ is proposed:

$$J = \sqrt{\frac{1}{n-1}\sum_{i=1}^{n}(\bar{T}_{\text{н}i}(Q,v,t_1) - \bar{T}_{\text{н}})^2} \rightarrow \min_{Q,v,t_1} \quad (1)$$

where $\bar{T}_{\text{н}i}(Q,v,t_1) = \frac{T_i(Q,v,t_1)}{T_{max}}$;

n is a sample in the fixed volume of heated metal.

The criterion of optimality is the minimum of the functional (1) in the fixed volume of heated metal with a variation of the parameter for which the solution is being searched.

From a physical point of view, the minimum of the proposed functional is proportional to the minimum scattering of the heating temperature in the fixed volume of metal. If we introduce volume restrictions on this function, then we can assume the following: in the case of uniform heating of the material, i.e. minimum temperature scattering in the volume under consideration, the probability of uniform melting of metal will be the highest. This will entail the formation of the seam with the best quality indicators. The introduction of normalized indicators increases the sensitivity of the criterion, which is especially important for temperatures not exceeding the temperatures of the phase transitions of the welded materials.

The calculation of the functional is based on the classical theory developed by N.N. Rykalin in the middle of the 20th century, suitable for analysing the temperature fields in bodies when they are heated with various sources of heat. In the derivation of expressions describing the temperature fields of mobile sources, we use the principle of superposition [23].

$$T_i(Q,v,t_1) = T_1(Q,v,t_1) + T_2(Q,v,t_1) \quad (2)$$

The representation of the heating source in the form of a sum of two instantaneous moving sources (2): point ($T_1(Q,v,t_1)$) and linear ($T_2(Q,v,t_1)$), is due to the need to take into account the effect of the product thickness on the thermal field.

The chosen formulas depend not only on the characteristics of the material being welded, but also on the electron beam energy parameters. Instant source models are valid for metals and alloys, provided that the physical characteristics of the material do not change when heated. To this end, as the initial temperature of the search for the integration time and the limits of integration of the thermal models we take the one to which the numerical values of the thermo-physical coefficients of the alloy correspond. For example, for aluminium alloy AMG-6, the thermal conductivity coefficient $\lambda = 1{,}17\frac{W}{cm\cdot K}$, and the volumetric heat capacity $c\rho = 2{,}43\frac{J}{cm^3\cdot K}$ are determined for the temperature $100°C$.

The thermal field generated by the heating of the material is by nature not stationary. Therefore, in order to prove the proposed functional to be an optimization criterion, it is necessary to determine the integration time, the parameters of the fixed volume of the metal, and the integration step for calculating the thermal field when the material is heated with the selected complex energy source.

It is important to choose the step in calculating the integrals in the models of the thermal field. Studies have shown that in order to find the region of variation of the investigated parameters, where the functional undergoes extreme value, the integration step is chosen so that the lines of the graphs are smoothed out. Thus, for example, for the AMG-6 alloy, the integration step was selected as $\Delta = 0{,}0005cm$. In the future, with refined calculations, in order for the graphs of calculated dependences to be uniform, the integration step had to be reduced to $\Delta = 0{,}00025cm$. This greatly increased the computing resource, however, it increased the accuracy of calculations.

Based on the results of long-term studies of the functional behaviour of various materials and their thicknesses, the authors developed an algorithm for computing the parameters of technological process, which are recommended for the future use in electron-beam welding.

4. Experimental study

At the first stage, the integration time is searched. To do this, we limit the heating temperature for the AMG-6 alloy to $100°C$ and compute the values of the functional (1) for the selected heating source. As a result, we obtain a dependence graph of the functional for the integration time (Fig. 1). On the graph, we determine the time coordinate in which the functional undergoes the extreme value (minimum); this time is subsequently taken for calculations.

IOP Conf. Series: Materials Science and Engineering **467** (2019) 012013 doi:10.1088/1757-899X/467/1/012013

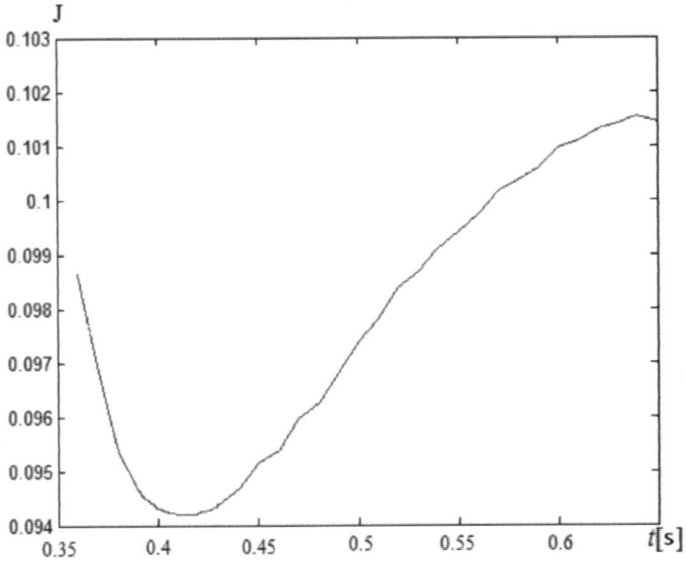

Figure1. The dependence plot of the functional J on the integration time t [c]

This graph is a curve constructed for the fixed value of energy of the heating source Q. The analysis of families of such curves constructed for different Q sources allowed us to localize those energy values for which the optimization criterion of the chosen functional was apparent. For this purpose, the slopes of the right branch of the curve were compared (Fig. 1).

The graph of coefficients is presented in Fig.2.

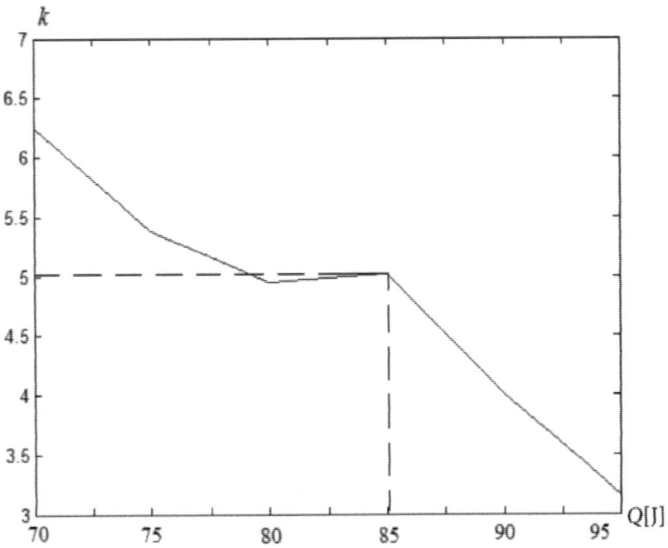

Figure2. The dependence plot of the slope of the curve's right branch on the energy Q source (for a plate 0.12 cm thick)

The graph (Fig. 2) is constructed for a plate 0.12 cm thick with the AMG-6 material. In the range of the energy source $Q = 70 \dots 95\,J$, the energy value $Q = 85\,J$ was determined, for which the selected optimization criterion is effective. For an accelerating voltage of 30 kV, this energy corresponds to welding amperage of 13 mA.

There can be more than one local extremum since the energy of the source ensuring effective heating of the metal is not always known. Therefore, when specifying a sufficiently wide range of energy changes in the calculations, there may be several extrema on the graph. For example, when calculating for a plate with a

IOP Conf. Series: Materials Science and Engineering **467** (2019) 012013 doi:10.1088/1757-899X/467/1/012013

thickness of 1.2 cm, there were two local extrema (Fig. 3). However, further calculations showed that not all local regions have solutions for the chosen optimization criterion. For a plate 1.2 cm thick, the functional became a minimum at the extremum point only in the energy region of the source $Q = 700\ J$. And in the field $Q = 840\ J$ the functional changed monotonically.

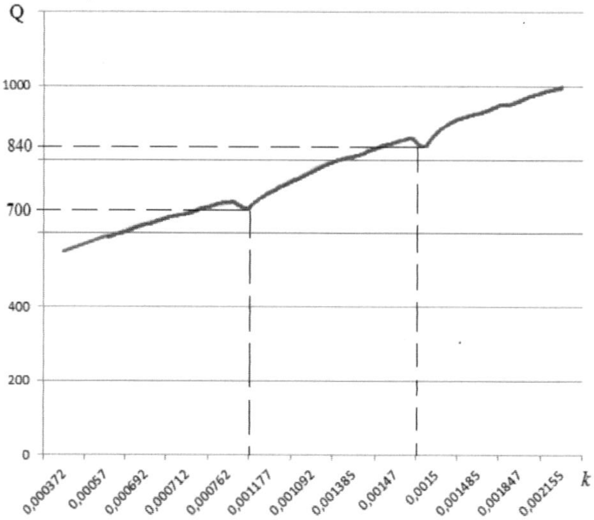

Figure 3. The plot of energy dependence of the Q source on the slope of the right branch of the curve *k* (for a plate 1.2 cm thick)

From the energy parameter *Q* found it is not difficult to calculate the necessary values of accelerating voltage *U* and welding amperage *I* using the formula:

$$Q = 0{,}24 \cdot U \cdot I \cdot \eta \tag{3}$$

where η is the efficiency (for EBW $\eta \approx 0{,}89$).

The second stage includes the search for temperature threshold that specifies a fixed volume in the material that ensures manifestation of the functional properties as an optimization criterion. Using the integration time obtained at the first stage, we plot the dependence of the functional on temperature (Fig. 4). On the graph, we determine the coordinate of the threshold temperature, in which the functional undergoes the extreme value (minimum). At this stage, we also calculate the width of the limited heating zone, which is equivalent to the width of the welded seam. This parameter can be used to predict the width of the seam for the case of changing technological welding modes, such as welding speed, welding amperage and change in focal length.

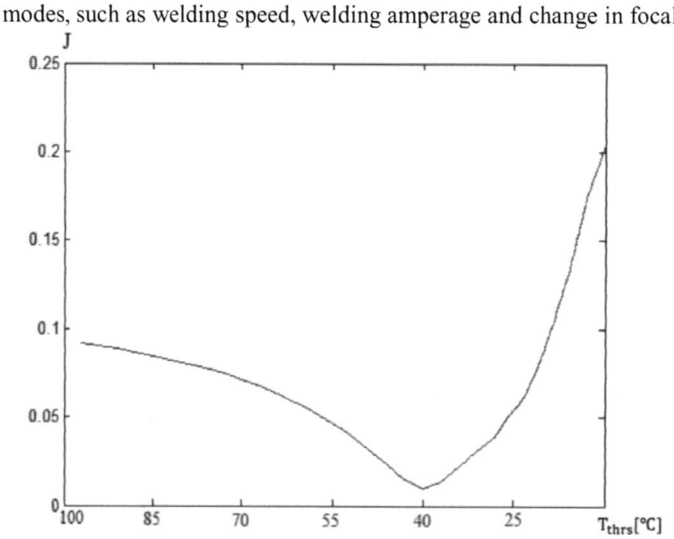

Figure 4. The plot of the functional *J* dependence on temperature threshold T_{thrs}[°C]

For a greater manifestation of the extreme properties of the functional at the first and second stages of the calculation, at each step of integration the variation of welding speed in the realizable range of the investigated material was used: $tv = 0{,}1 \ldots 3{,}0\ ^{cm}/_{s}$.

Having obtained the integration time and boundaries for calculating the welding parameters using the extreme properties of the functional, we proceed to direct determination of these parameters. Figure 5 shows the graph of the functional dependence on the welding speed. On the graph, we determine the velocity coordinate, at which the functional undergoes the extreme value (minimum).

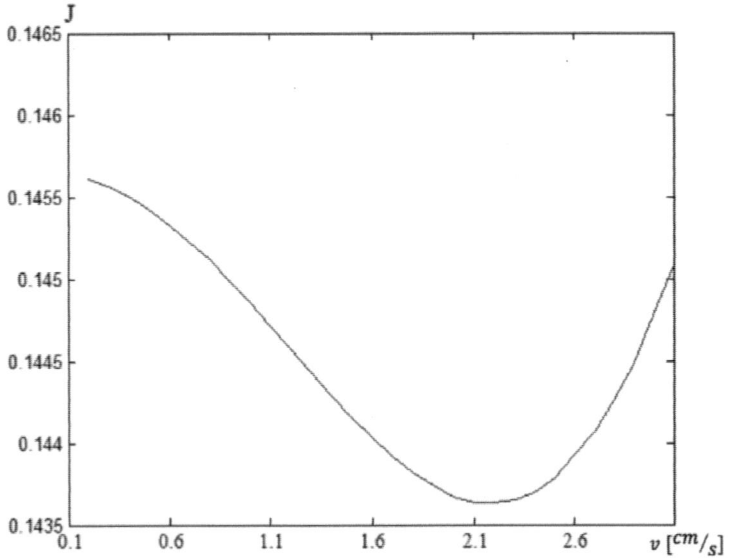

Figure 5. The dependence of the functional *J* on the welding speed $v[^{cm}/_{s}]$

By performing analogous calculations and changing numerical energy values of heat sources Q, we can consolidate the extrema obtained in the calculations in the form of amplitude dependence graphs of the functional *J* extrema (Fig. 6) for an AMG-6 alloy with the thickness of 0.12 cm.

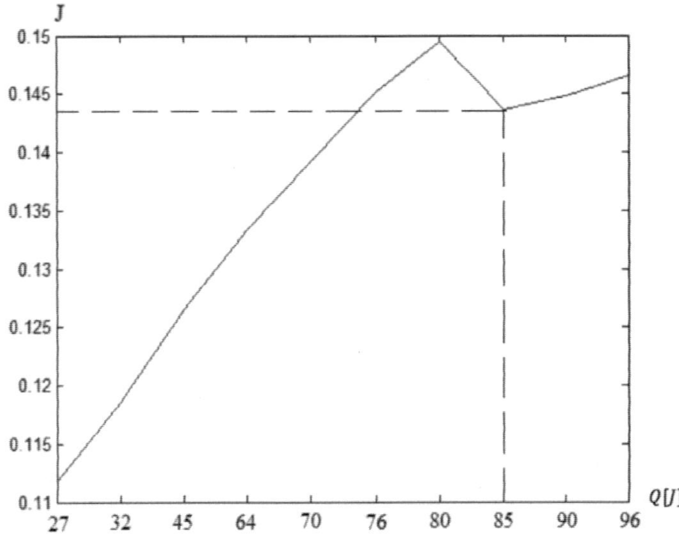

Figure 6. The dependence graph of the functional *J* on the energy source $Q[J]$

These three graphs are of the greatest practical interest when choosing welding modes for a particular material with a fixed penetration depth or the corresponding thickness of the welded plate.

The extremum of the functional is obtained only for fixed parameters of variation intervals of coordinates x and y, which are characteristic for a particular thickness and the material to be welded. Currently, studies are under way to develop a search algorithm for these quantities.

Thus, it can be assumed that the welding speed for products of other thicknesses or in the case of complex configurations of joints can be selected using the calculation method presented in this paper. The application of the theory of the thermal field and the criterion of a minimum of the functional proposed by the authors allows us to calculate not only the welding speed and optimize the power of the electron beam, but also to select these parameters according to the expected width of the weld for products of various materials and their combinations (bimetals), and in the process of welding only adjust this parameter.

The authors also proposed a method for modelling the effect of the focal plane position of the electron beam on the thermal process relative to the heated material surface. The effect of the focus of the electron beam on the thermal process of heating can be modelled if the instantaneous point source T_1 along with the linear source T_2 is moved along the entire depth of metal z_f in the selected combined energy source. The graph of dependence of the functional J on the depth of the electron beam penetration z_f is shown in Fig. 7.

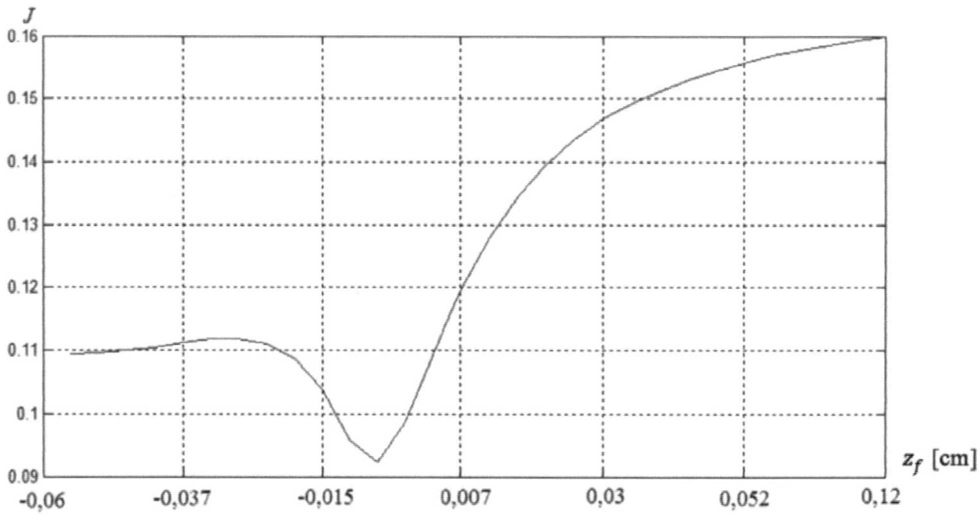

Figure 7. The dependence graph of the functional J on the depth of the electron beam penetration z_f [cm]

In case with a work piece thickness of 0.12 cm, the minimum value of the functional indicates the practical coincidence of the focal spot of the electron beam with the surface of the part.

Calculations showed that when the amperage and the welding speed change on the part with the same thickness, the position of the focal spot does not change.

When modelling a thermal heating process, the calculations used characteristics and test results of the samples obtained with electron beam equipment with an accelerating voltage of 30 kV.

5. Conclusion

Application of the optimality criterion, taking into account the uniformity of heating the weld pool, allows to determine the main parameters of the technological mode of electron beam welding: the welding amperage, focal length and welding speed.

Numerical modelling of thermal processes to determine EBW parameters will significantly reduce the cost of technology development for structures from new materials.

The reported study was funded by Russian Foundation for Basic Research, Government of Krasnoyarsk Territory, Krasnoyarsk Region Science and Technology Support Fund to the research project № 17-48-240098.

References

[1] Rykalin N., Zuev I. & Uglov A., 1978 Fundamentals of electron-beam processing of materials 239
[2] Schiller Z., Gaizig U. & Panzer Z., 1980 Electron-beam technology. Moscow: Energia 528
[3] Nazarenko O., Kaidalov A. & Kovbasenko S., 1987 Electron-beam welding. Kyiv: Naukova Dumka, 256

[4] Orlinov B 1982 Electronic and ionic methods and devices for processing and analyzing the substance
[5] Mladenov G 2009 Electronic and ion technologies. Acad. M. Drinov, Sofia 387.
[6] Dvorkin I Ledovskoy V & Mladenov G 1972 Electron-beam welding of thin-walled parts in a continuous mode. Moscow: Electronic engineering 53-59
[7] Petrov P & Mladenov G 1991 Theoretical analysis of heat flow and structural changes during electron beam irradiation of steel. Vacuum **42(1-2)** 29-32
[8] Koleva E & Vuchkov I 2005 Model-based approach for quality improvement of electron beam welding applications in mass production. Vacuum **77(4)** 423-428
[9] Rykalin N 1985 Laser and electron-beam processing of materials: reference book. Moscow.: Mechanical engineering 496
[10] Goldak J Chakravarti A & Bibby M 1984 A new finite element model for welding heat sources. Metallurgical and Materials Transactions B **15(2)** 299-305
[11] Nguyen N 2004 Thermal analysis of welds. Applied Mechanics Reviews, 57, B34
[12] Nguyen N T, Mai Y W, Simpson S & Ohta A 2004 Analytical Approximate Solution for Double Ellipsoidal Heat Source in Finite Thick Plate. Welding Journal 83(3) 82
[13] Sabapathy P N, Wahab M A & Painter M J 2001 Numerical methods to predict failure during the in-service welding of gas pipelines. The Journal of Strain Analysis for Engineering Design **36(6)** 611-619
[14] Zhang J & Dong Y 2001 U.S. Patent No. 6,324,491. Washington, DC: U.S. Patent and Trademark Office
[15] Goldak J, Bibby M, Moore J, House R & Patel B 1986 Computer modeling of heat flow in welds. Metallurgical and Materials Transactions B **17(3)** 587-600
[16] Ranatowski E & Pocwiardowski A 1998 An analytic-numerical evaluation of the thermal cycle in the HAZ during welding. BOOK-INSTITUTE OF MATERIALS **695** 379-395
[17] Karkhin V A, Plochikhine V V, Ilyin A S & Bergmann H W 2002 Inverse modelling of fusion welding processes. Welding in the World **46(11-12)** 2-13
[18] Rayamyaki P, Karkhin V A & Khomich P N 2007 Determination of the main characteristics of the temperature field for the evaluation of the type of solidification of weld metal in fusion welding. Welding International 21(8) 600-604
[19] Karkhin V A, Khomich P N, Ossenbrink R & Mikhailov V G 2007 Calculation-experimental method for the determination of the temperature field in laser welding. Welding International **21(5)** 387-390
[20] Dey V, Pratihar D K, Datta G L, Jha M N, Saha T K & Bapat, A V 2009 Optimization of bead geometry in electron beam welding using a Genetic Algorithm. journal of materials processing technology **209(3)** 1151-1157
[21] Rouquette S, Guo J & Le Masson P 2007 Estimation of the parameters of a Gaussian heat source by the Levenberg–Marquardt method: application to the electron beam welding. International Journal of Thermal Sciences **46(2)** 128-138
[22] Wentzel E 1988 Operations Research: Tasks, Principles, Methodology
[23] Zuev I, Rykalin N & Uglov A 1976 Effective efficiency electron-beam heating. Physics and Chemistry of Materials Processing **(1)** 22-25

IOP Conf. Series: Materials Science and Engineering **467** (2019) 012014 doi:10.1088/1757-899X/467/1/012014

Influence of cation substitution on dielectric and electric properties of bismuth stannates$Bi_2Sn_{1.9}Me_{0.1}O_7$ (Me=Cr, Mn)

S S Aplesnin[1,2], L V Udod[1,2], Y Y Loginov[1], V V Kretinin[1], A N Masyugin[1]

[1]Reshetnev Siberian State University of Science and Technology,660037, Krasnoyarsk, Russia
[2]Federal Research Center "Krasnoyarsk Science Center" of the Siberian Branch of the Russian Academy of Sciences660036, Krasnoyarsk, Russia

E-mail: apl@iph.krasn.ru

Abstract. The article studies effect of nonstoichiometric substitution of the tin ions by chromium and manganese ions on the dielectric and electrical properties of bismuth pyrostannate. The research performs measurements of the current-voltage characteristics, capacitance and tangent of the dielectric loss angle in the temperature range from 300 to 800 K; it finds a qualitative difference in the temperature behavior of the permittivity of bismuth pyrostannate with chromium and manganese ions. A change in the type of conductivity from hopping to tunneling emission is established. The change in the activation energy as a function of the ion radius is found.

1. Introduction
Bismuth pyrostannate attracts attention both from the fundamental point of view as multiferroics, spin ice [1] and as optical devices [2] and sensors [3] use it. Substitution of the tin ions in $Bi_2Sn_2O_7$ by 3d magnetoactive ions will lead to a magnetic order that can be controlled by an electric field as a result of the displacement of the oxygen ion. Variation of ferroelectric properties by a magnetic field is due to magnetoelastic interaction. Ferroelectric properties may be enhanced by non-stoichiometric substitution which gives an additional electron contribution to the polarization.

The aim of this study is to establish the role of non-stoichiometric substitution of tin ions by chromium ions as compared to manganese ions on the dielectric and electrical properties of bismuth pyrostannate.

2. Experimental
Samples of Bi_2 $(Sn_{0.9}Me_{0.1})$ $_2O_7$(Me = Cr,Mn) were synthesized using the solid-state reaction technique. The powder X-ray diffraction pattern of samples was obtained at room temperature using a Bruker D8 ADVANCE diffractometer, a VANTEC linear detector, and the CuK_α - radiation. All peaks, except several weak impurity peaks of an unknown phase corresponded, to the monoclinic *Pc* cell in the α-phase of $Bi_2Sn_2O_7$ [4]. The crystal structure contains 32 Bi^{3+} ions, 32 Sn^{4+} ions and 112 O^{2-}ions in the independent part of the cell. Fig.1 shows the crystal structure of $Bi_2(Sn_{0.9}Mn_{0.1})_2O_7$. All Bi^{3+} ions are surrounded by eight O^{2-} ions and form a distorted cubeand Sn^{4+} is located in the oxygen octahedron, which are connected by vertices.

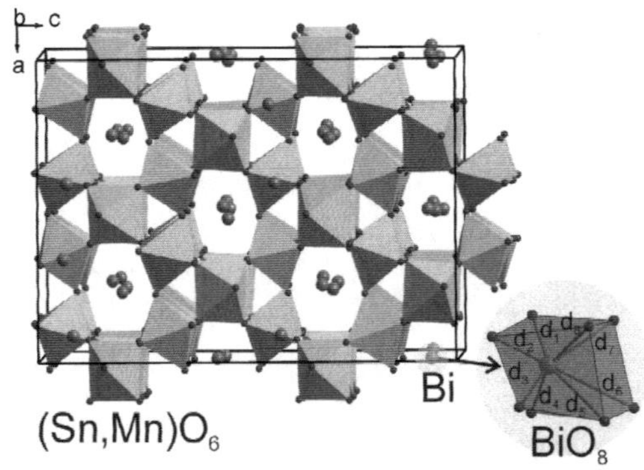

Figure 1. Crystal structure of $Bi_2(Sn_{0.9}Mn_{0.1})_2O_7$. The inset shows one of 32 independent polyhedrons BiO_8 that form the unit cell. The d_{1-4} bond lengths takes values 2.2–2.5 Å, and the d_{5-8} bond lengths takes values 2.5–3.2 Å

The decrease in the cell volume $Bi_2 (Sn_{0.9}Me_{0.1})_2O_7$ (Me = Cr, Mn) correlates with the ionic radius IR (Cr^{3+}) = 0.615 Å, IR (Mn^{4+}) = 0.53 Å, they are smaller than the tin ion radius IR (Sn^{4+}) = 0.69 Å [5].

The electrical properties of $Bi_2 (Sn_{0.9}Me_{0.1})_2O_7$ (Me = Cr, Mn) are investigated by a four-probe technique on a 6517 V electrometer in the temperature range of 350–700 K. Capacitance, and loss-angle tangent are measured on an AM-3028 component analyzer in the frequency range of 0.1–1000 kHz at temperatures of 300–750 K. The current-voltage characteristics are obtained on an electrometer Keithley 6517b in an electric field from -800 to 800 V/cm.

3. ResultsandDiscussion

Permittivity. Figure 2 shows the real and imaginary parts of the $Bi_2(Sn_{0.9}Cr_{0.1})_2O_7$ permittivity versus temperature for several frequencies. The temperature dependencies of ε(T) reveal inflection points in the real part (Fig. 2a) and wide maxima of the dielectric losses (Fig. 2b) at the same temperature, increasing from T_m = 370 K to 560 K according to the logarithmic law from frequency. The relaxation time is described by the exponential law$\tau=\tau_0 epx(\Delta E/kT)$ with an activation energy ΔE = 0.4 eV. The dielectric susceptibility increases sharply at approaching to the $\beta \rightarrow \gamma$ phase transition temperature.

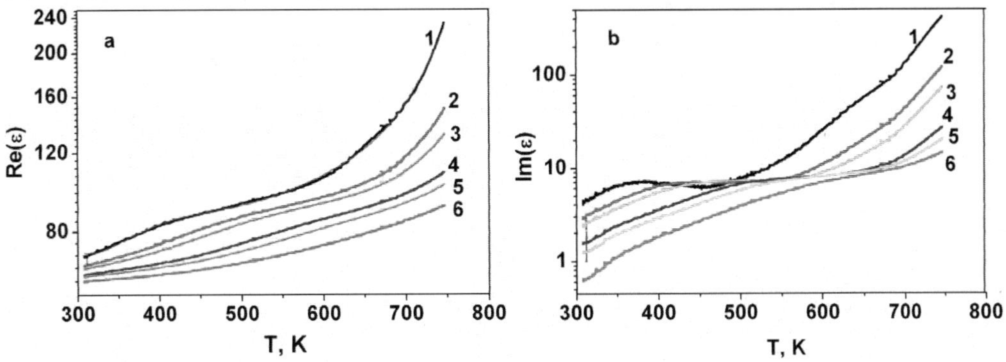

Figure 2. Temperature dependence of the permittivity of $Bi_2(Sn_{0.9}Cr_{0.1})_2O_7$ at different frequencies: (a) real part and (b) imaginary part. Curve 1 corresponds to 1 kHz, 2 - 5 kHz, 3 - 10 kHz, 4 - 50 kHz, 5 - 100 kHz, 6 - 300 kHz

The permittivity in the range of $5\cdot10^4$-10^6Hz frequencies and in the range of 300-500 K temperatures is well described in term of the Debye model:

$$Re(\varepsilon)=\varepsilon_0+\chi_0/(1+(\omega\tau)^2), \quad (1)$$

$$Im(\varepsilon)=\chi_0\omega\tau/(1+(\omega\tau)^2), \quad (2)$$

where ε_0 is the temperature-independent contribution to the permittivity, χ_0 is the static susceptibility of the dipoles, ω is the frequency, and τ is the relaxation time of $1.5\cdot10^{-5}$sec. At temperatures above 550 K the permittivity is described by a power function $\varepsilon=A/\omega^n$with n <1. The exponent increases at the heating.

The real and imaginary parts of the permittivity are well described by the exponential law lnε = A + ΔE/T with the activation energy ΔE = 0.3 eV for the real part and ΔE=0.6 eV for the imaginary part of the permittivity. The activation process is due to the diffusion of current carriers in the domain and the accumulation of charge on the surface of the domain.

The real part of the permittivity $Bi_2(Sn_{0.9}Mn_{0.1})_2O_7$(Fig. 3a) has a small minimum in the vicinity of $\alpha \rightarrow \beta$ phase transition. At farther heating the permittivity sharply increases. The anomalytemperature isagreement with the temperature of small maximum on the differential-scanning calorimetry curve at T = 420 K [6]. The real and imaginary parts of the permittivity reveal a wide maximum at T>500 K, that is absent on the DSC curve. A similar temperature dependence of the permittivity above T = 550 K is observed in $Bi_2(Sn_{0.9}Cr_{0.1})_2O_7$.

The behavior of the permittivity at high temperatures occurs due to a change in the electronic structure as a result of the large electronic polarization of the Bi-O bond. Oxygen ions are known to possess a significant polarization; it increases at the presence of covalence and anisotropy. The polarization of bismuth ions arises from $6s^2$ lone pair electrons. The effect of the lone pair causes a large mobility of oxygen ion in the Bi_4O' sublattice and strong displacements of them from the sites of the ideal lattice of the pyrochlore structure. The absence of significant structural changes in $Bi_2Sn_2O_7$ in the temperature range of 500-600 K; it allows to make the assumption that the observed singularities in the behavior of the dielectric susceptibility due to processes of charge transfer and a sharp change in the ions polarization. A similar behavior of the dielectric constant is observed in α-Bi_2O_3 [7]. The substitution of Sn ions by Cr and Mn ions will lead to deformation of the oxygen - bismuth bond and to an increase in the dipole moment, since their ionic radii are different. The heterovalent substitution of tin by chromium leads to an electronic contribution to the permittivity and an increase in Re (ε) compared to the manganese ions giving the ionic contribution.

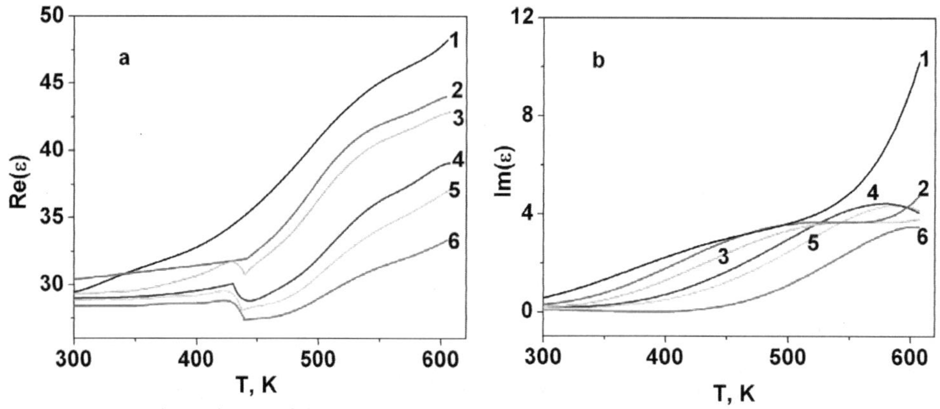

Figure 3. Temperature dependence of the permittivity of $Bi_2(Sn_{0.9}Mn_{0.1})_2O_7$ at different frequencies: (a) real part and (b) imaginary part. Curve 1 corresponds to 1 kHz, 2 - 5 kHz, 3 - 10 kHz, 4 - 50 kHz, 5 - 100 kHz, 6 - 300 kHz

Electrical properties. The electrical resistance determined from the current-voltage characteristic shows a typical semiconductor conductivity for two samples of $Bi_2(Sn_{0.9}Me_{0.1})_2O_7$(Me=Mn, Cr). The electrical resistance for $Bi_2(Sn_{0.9}Cr_{0.1})_2O_7$ is independent on temperature at T=500-550 K, that agrees with the anomalies in the dielectric susceptibility. The change from the Mott conductivity mechanism to the Poole-Frenkel conduction mechanism occurs also in this temperature range. The values of activation energies ΔE, calculated from the temperature dependence of lg(ρ) on the reciprocal temperature (Fig.4) are described by the linear function $ln\rho=ln\rho_0+\Delta E/(k_BT)$ in the temperature range 300-500 and are equal to 1.58 and 1.17 eVfor $Bi_2(Sn_{0.9}Mn_{0.1})_2O_7$ and for $Bi_2(Sn_{0.9}Cr_{0.1})_2O_7$.

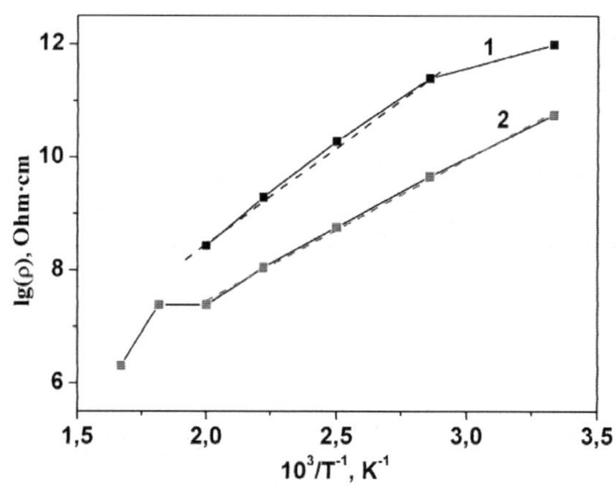

Figure 4. Temperature dependence of logarithm of the resistivity of $Bi_2(Sn_{0.9}Mn_{0.1})_2O_7$ and $Bi_2(Sn_{0.9}Cr_{0.1})_2O_7$. Curve 1– $Bi_2(Sn_{0.9}Mn_{0.1})_2O_7$, 2 – $Bi_2(Sn_{0.9}Cr_{0.1})_2O_7$

The conduction mechanism and the existence of inhomogeneous electronic states are determined from the current-voltage characteristics. The shape of the curve I (U) varies with increasing temperature. In the temperature range where charge ordering is observed, the conductivity is described in a model of currents limited by the space charge and is described by the quadratic Mott law [8] for both $Bi_2(Sn_{0.9}Mn_{0.1})_2O_7$ and $Bi_2(Sn_{0.9}Cr_{0.1})_2O_7$:

$$j=\frac{9}{8}\tau_{\mu}\sigma_0\mu\frac{U^2}{L^3}, \qquad (3)$$

where j – the current density, τ_μ– the Maxwell relaxation time, σ_0 – the electrical conductivity in the bulk of the material in the absence of charge carrier injection, μ – the mobility of the charge carriers, U – the applied voltage, and L – the thickness of the sample. In Fig. 5a, the experimental data of the I-V characteristic in logarithmic coordinates are well described by the relation (3) for $Bi_2(Sn_{0.9}Mn_{0.1})_2O_7$ to T = 400 K, and for $Bi_2(Sn_{0.9}Cr_{0.1})_2O_7$ to T=500 K. The Pool- Frenkel conduction mechanism prevails in the range of high temperatures at T>450 K for the $Bi_2(Sn_{0.9}Mn_{0.1})_2O_7$ and at T >550 K for $Bi_2(Sn_{0.9}Cr_{0.1})_2O_7$ (Fig. 5b), according to which the strong electric field applied to the sample changes the form of the potential barriers for charge carriers. In this case, the current depends exponentially on the applied voltage and the exponent contains the square root of voltage:

$$I = e\mu n_0 \frac{U}{L}\exp\frac{\beta U^{1/2}}{kTL^{1/2}}, \qquad (4)$$

where, e – the charge of the electron,μ – the mobility of the charge carriers, n_0- the electron concentration in the conduction band in the absence of the field, U – the applied voltage, L – the thickness of the sample, k – the Boltzmann constant,β – the Pool-Frenkel constant, $\beta = \left(\frac{e^3}{\pi\varepsilon\varepsilon_0}\right)^{1/2}$, ε is the permittivity of a semiconductorε_0 – the dielectric constant. The linearity plots in the ln(I/U) the dependence from $U^{1/2}$in the Pool-Frenkel coordinates indicates that the transport of charge carriers is carried out mainly by as hopping conduction mechanism and due to the tunneling electron emission [9].

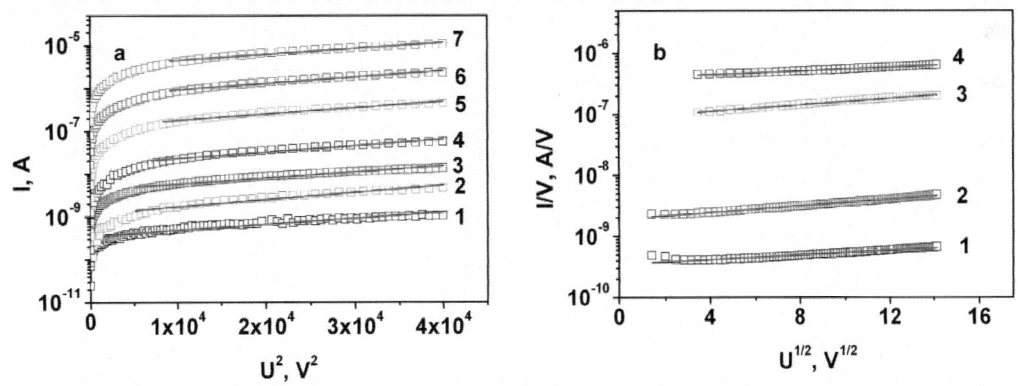

Figure 5.I–V characteristics of $Bi_2(Sn_{0.9}Mn_{0.1})_2O_7$ and $Bi_2(Sn_{0.9}Cr_{0.1})_2O_7$, positive voltage. (a) Dependence of the current on the squared applied voltage. Curve 1,3 corresponds to $Bi_2(Sn_{0.9}Mn_{0.1})_2O_7$ at T = 350 and 400 K, 2,4,6,7 - $Bi_2(Sn_{0.9}Cr_{0.1})_2O_7$, T = 300, 350, 400 450 and 500 K accordingly. (b) Pool–Frenkel I–V characteristic measured. Curves 1,2 - $Bi_2(Sn_{0.9}Mn_{0.1})_2O_7$ at T = 450 and 500 K; 3,4 - $Bi_2(Sn_{0.9}Cr_{0.1})_2O_7$, T = 550 and 600 K

4. Conclusion

The maximums in the temperature behavior of the dielectric loss at $Bi_2(Sn_{0.9}Cr_{0.1})_2O_7$ hole doping for α – phase are found. The jump in the temperature dependence of permittivity and an electronic transition are detected in the case of isovalent substitution for $Bi_2(Sn_{0.9}Mn_{0.1})_2O_7$. The electronic transition with the variation in the conductivity type from hopping to Pool–Frenkel emission is established. The research also determines a decrease in the activation energy at hole substitution as compared to the isovalent substitution tin ions by manganese.

This study was supported by the Russian Foundation for Basic Research project N 17-32-50080, the state order № 3.5743.2017/6.7.

References

[1] Yaraskavitch L R,Revell H M, Meng S,Ross K A,Noad H M L,Dabkowska H A, Gaulin B D,and Kycia J B, 2012 *Phys. Rev. B*,**85**, 020410(R)
[2] Janner D, Tulli D, Jofre M, Yudistira D, Balsamo S, Belmonte M, Pruneri V 2013 *IEEE J. Sel. Top. Quantum Electron,***19**, 3400610
[3] Tulli D, Janner D, Garcia-Granda M, Ricken R, Pruneri V 2011 *J. Appl. Phys. B,***103**, 399
[4] Radosavljevic Evans I, Howard J A K, Evans J S O 2003 *J. Mater. Chem.,***13**, 2098
[5] Shannon R D 1976 *ActaCryst. A,***32,** 751

[6] Aplesnin S S, Udod L V, Sitnikov M N, Molokeev M S, Tarasova L S, and Yanushkevich K I 2017 *Phys. Solid State,***59,** 2268
[7] Orlov V G, Bush A A, Ivanov S A, and Zhurov V V 1997 *Phys. Solid State*, **39,** 770
[8] Mott N F and Davis E F *Electronic Processes in Non-Crystalline Materials* (Oxford, 1971)
[9] Macdonald J R 1990 *ElectrochimicaActa,***35,** 1483

IOP Conf. Series: Materials Science and Engineering **467** (2019) 012015 doi:10.1088/1757-899X/467/1/012015

The 1939 Northern Sea Route expedition for the GULAG of the NKVD: a study of vessel ice protection

A E Goncharov
Associate professor, Reshetnev Siberian State University of Science and Technology, Krasnoyarsk, Russia

E-mail: goncharovae@sibsau.ru

Abstract. Arctic navigation has always been a challenge both to the mariner and his vessel. Before the epoch of icebreakers, many polar expeditions used conventional ships, which were often strengthened to cope with the ice-infested waters of the Arctic Ocean. Unfortunately, little is known on the design of this technology and its effectiveness. This paper aims to give some insights on this problem by describing the ice protection used during a Soviet Northern Sea Route expedition staged by the GULAG of the NKVD in 1939.

1. Introduction

The practice of sheathing ship hulls with metals (primarily copper) to protect the wood from accumulating marine life has a long history. As a massive technical innovation in shipbuilding this technique arrived only during the closing decades of the eighteenth century [1]. Along with being an effective means of protection against biological fouling, sheathing had another important application – hull protection against polar ice. And, as the age of wooden ships gradually drew to an end, sheathing became associated not with the copper-bottoms of Nelson's navy, but with ice protection for polar vessels (see, for example, the meaning of *sheathing* in the dictionary of Newfoundland English [2]).

The evolution of polar ship design has largely been limited to the history of icebreakers with researchers occasionally enhancing their paper by adding some passages on the history of polar shipbuilding before the epoch of icebreakers. These additions, however, are often limited to the description of specialized polar vessels such as Fridtjof Nansen's *Fram* [3], ignoring conventional ships that were deliberately strengthened for operating in icy waters. Until the 1950's and the epoch of intense icebreaker construction, the main vehicles of the busiest Arctic marine highway, the Northern Sea Route (NSR), were conventional vessels. In some cases these ships were equipped with limited means of ice protection. This is especially true for the vessels of the voyages that pioneered the NSR in the second half of the nineteenth century, and the annual expeditions organized during the Soviet era (especially the Kara Sea (Barter) Expeditions of the 1920's).

Arctic expeditions of the Industrial Revolution had their ships covered with so-called 'ice belts' which protected those parts of the hull which were exposed to the ice and, hence, could be damaged. Unfortunately, there is very little knowledge on the way this specific protection was installed, the materials used for its production, and how effective the ice belts were. For instance, Frederick William Beechey the commander of the British polar expedition aboard the HMS *Blossom*, modestly mentions that his ship (which sailed in the Chukchi Sea in 1826) was "partially strengthened, and other wise adapted to the service" [4, p. iv]. We see the same situation on the pages of narratives, written by other polar explores of the age [5 p.; 6, p. iii].

There is much more knowledge on the construction of the ice belts on the HMS *Erebus* and *Terror* of Sir John Franklin's Arctic expedition, mainly because of the intense discussion around this expedition due to its tragic fate. Significant interest among scholarship was generated around the ships' design after the discovery of both ships in 2014 and 2016. From topical publications, we learn that some of the reinforcements to the ships' hulls included diagonally cross-planked decks, a second layer of planking, which doubled the hull thickness, watertight bulkheads, and iron casing on the bows [7].

Another example of strengthening a wooden vessel for polar navigation during the Franklin era is that of *Fox*, the steam yacht used during Francis Leopold McClintock's Arctic search expedition of 1857. This is how McClintock describes the process of adapting *Fox* to Arctic navigation:

'... the whole vessel had to be externally sheathed with stout planking, and internally fortified by strong cross beams, longitudinal beams, iron stanchions, and diagonal fastenings; the false keel taken off, the slender brass propeller replaced by a massive iron one , the boiler taken out, altered, and enlarged; the

sharp stem to be cased in iron until it resembled a ponderous chisel set up edgeways ...' [8, p. 6].

These two schemes illustrates the basic concept for protecting polar vessels, which consisted of the following modifications: (1) adding protection to the hull by increasing the its thickness and partially casing the hull with metal to reduce skin friction; (2) adding framing to transverse stiffen the ship (in some cases longitudinal stiffing may have also been applied).

None of these vessels were, of course, designed to cope with heavy ice conditions and withstand the immense force exerted onto their hulls by shifting ice fields. The conversion of former military and civil vessels into polar ships continued through the nineteenth and early twentieth century. Notable examples include the *Philomel*-Class *Jeannette* and *Blencathra* (the second vessel was later renamed to *Sviataia Anna*). *Blencathra* was even equipped with such a piece of technology as an ice ram [9], the construction of which is not specified in available historical sources. Most likely, the ram was a casing of iron on the ship's bow, which protected it from impact with ice (likely, similar to the stem of *Fox*). The workhorse of polar expeditions during the first decades of the twentieth century was the Dundee whaler and other whalers, which operated in the polar seas of both hemispheres.

Having briefly discussed the milestones of polar ship protection development to the twentieth century, we can now proceed to investigating an episode in the history of designing and exploiting ice protection for the hulls of conventional vessels. A method for constructing such a protection was developed by Soviet naval engineer Andrei Ivanovich Dubravin (1898–1988). While he remains largely an unknown figure in the history of the Northern Sea Route (especially for the Anglophone reader), Dubravin played an important role in the development of icebreaking technology in the Soviet Union. He is best known by his book on the 1929–1930 Kolyma expedition aboard the steamer *Stavropol'* [10]. During this expedition, Dubravin (then, an engineering student) performed a series of important investigations on the behavior of ship hulls in ice, greatly contributing to this area of materials science and naval engineering. In 1939, Dubravin successfully tested his ice protection during a transit Northern Sea Route Expedition. His design was used for protecting various vessels traversing the sea route (one episode even includes covering the Soviet Navy *Shchuka*-class submarine *Shch-423* with a wood and metal sheath for the voyage [11]). However, despite its impact on Soviet Arctic navigation, the architecture of metal-and-wooden ice protection has not yet been thoroughly discussed in scientific literature. At the same time, if we give a detailed description of the ice protection, another question inevitably arises, "How efficient was this ice protection?" Considering this, we shall address these two questions in this paper.

2. The 1939 NKVD expedition. During the 1930's, the industrialization of the Soviet Far East and the northeastern part of Siberia was well under way. It was based primarily on the extraction of natural resources (especially minerals), which were vital both to the industrial plants of European Russia and to Soviet international trade. However, despite the abundance of many raw materials in Siberia and the Far East, their development was often impossible due to the absence of effective ways of communication to these remote territories. Not surprisingly, the Soviet government invested greatly into such massive projects as the Northern Sea Route and developing river navigation on all the major rivers of the Asiatic part of the country. Such were the settings for the polar shipping expedition of 1939.

In March 1939, the People's Commissariat of Domestic Affairs (the NKVD) arranged the delivery of cargo vessels and river dredges to the GULAG facilities in the Far East (the port of Nikolaevsk on the River Amur) [12, l. 121]. The vessels had been purchased in Western Europe and included one suction dredge and three bucket ladder dredges. The latter were identical and had the following technical characteristics: (1) the vessels had one deck; (2) the vessels did not have double bottoms; (3) the stern of each of the vessels had a slit to accommodate the dredging machinery, having, thus, a double hull (there were two rudders and screws, which made the transportation of the vessel through ice extremely difficult); (4) the vessels had low freeboards; (5) the vessels had tall superstructures; this had negative effect stability. The table below contains the dimensions of the three *Lena*-type dredges (*Lena, Pechora, Msta*) and the suction dredge *Zeia* [12, l. 74–76].

Table 1. General characteristics of dredges.

Vessel	Maximum length	Maximum breadth	Board height	Maximum draft	Displacement
Lena-type dredges	73.2 m	12.9 m	5.7 m	4 m	3,000 tons
Zeia suction dredge	67.6 m	16 m	5.33	4 m	3,350 tons

IOP Conf. Series: Materials Science and Engineering **467** (2019) 012015 doi:10.1088/1757-899X/467/1/012015

As the dredges were extremely expensive, special measures had to be taken to make them as ice resistant as possible. While such techniques as electric arc welding were comparatively novel, the overall concept of building a special ice protection for the ships' hulls, basically, repeated designs used in earlier polar vessels. A special commission consisting of academics, naval engineers, and NKVD authorities gathered to decide whether it was safe to transport these vessels from Leningrad to the Far East via the Northern Sea Route [12, l. 121–121verso]. However, the southern route through the Panama Canal was considered even more dangerous for the shallow-draft vessels. Moreover, with the approaching world war, the commission decided that it was much safer (and cheaper) to deliver the dredges by way of the Northern Sea Route. This decision triggered immediate actions for providing the ships with satisfactory protection against the ice; the option of staging the expedition without properly fitting the vessels was not even discussed. For instance, the maritime authorities of the Baltic Steam Shipping Company concluded that without providing the vessels with proper ice protection for the hulls and additional structural reinforcement, they cannot be taken to the Arctic [12, l. 98]. It was also suggested that each ship should be equipped with at least two motor-driven water pumps. By the time the ships were ready for the voyage, this list of safety equipment had been significantly expanded. Each vessel was equipped with: (1) electric arc welding equipment (including welding rods for working underwater); (2) spare steel sheets, girders, bolts, pins, and other fasteners; (3) one motorized marine pump with a capacity of 40 tons/hour. Besides the individual equipment for each vessel, the expedition carried two powerful (250–300-ton/hour) marine pumps and diving gear (including underwater lighting and telephone systems) [12, l. 90].

3. Constructing the ice protection.

The most important part of the project was, of course, the ice protection (Figs. 1, 2). Its design is exhaustively described by Dubravin.

A strip of steel 5 × 60 mm was welded along the length of each frame. Threaded pins having a diameter of 5/8 inch were welded perpendicular onto the strip. The first layer of wood was then mounted onto these pins. The layer consisted out of planks having a thickness of 90 mm, width of 120 – 200 mm (depending on the vessel), and a length of 6 m. The nuts and washers were sunk into the wood so that another layer of planking could be added. It consisted of thinner planks (65–75 mm) with the same width and length as those of the first layer. The planks were attached to the first layer using wire nails. Finally, a layer of metal was applied to protect the wood. The iron sheets (plates) were attached to the planking with rag bolts. The sheets overlapped one another and the joints between them were electric arc welded (a continuous lap weld with a pitch spacing of 60 × 120 mm). The iron sheets were overlapped in such a manner that when the ship moved forwards, the sheets would cover the joints offering minimum skin friction.

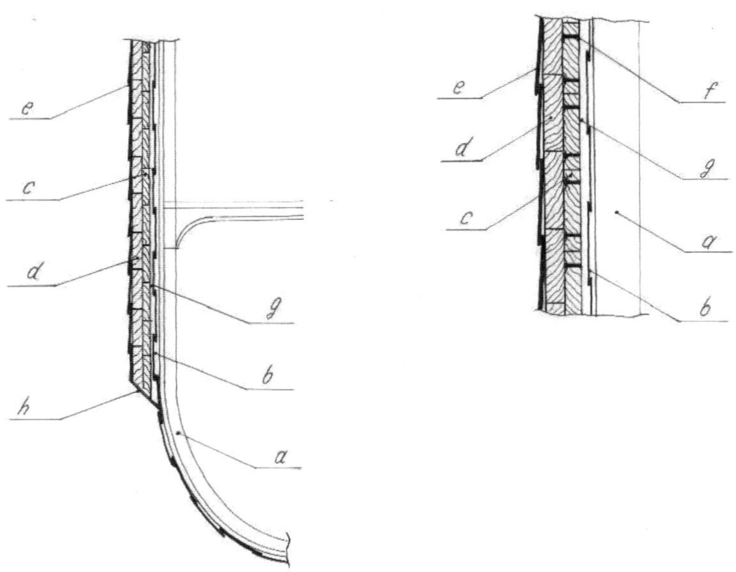

Fig. 1. Cross-section view of sheathing (a – ship frame; b – hull plate; c – first timber layer; d – second timber layer; e – metal sheathing (outer) layer); f – threaded pin; g – steel strip).

The protection was extended below the waterline almost to the chine; here the planks and metal structures were rounded so that the entire construction had a streamlined design eliminating any protruding components. This curvature was also covered by iron sheets 2–3 mm thick, which were welded both to the iron sheets of the ice layer and the chine hull of the vessel using intermitted seam welds.

The sheathing covered only those parts of the hull that were at risk of being damaged by ice; therefore emphasis was made on protecting the bow. Here the ice layer was extended almost to the deck [12, l. 81 – 83].

The vessel hulls were also strengthened from the inside in order for them to be able to resist the immense pressure of the ice in case they are beset in ice. Thus, according to Dubravin's description on the reinforcing structure, the longitudinal elements (or stringers) were additionally joined to an additional set of strengthened frames. Each frame had a wall height of 500 mm and a protruding band with a width of 120 mm and a thickness of 12 mm. The frames and stringers were joined using steel knees 700 long, 400 mm wide, and 10 mm thick. The whole structure was then secured using 100-mm-long fillet welds having a pitch length of 100 mm. The frames were welded to the floor and the cross-tie beams with a continuous fillet weld (throat gauge 7–10 mm). In addition, the forepeaks of the vessels were reinforced by timber cribs built between the floors and frames [12, l. 84].

Fig. 2. Assembly of sheathing on one of the dredges. Source: RGAVMF file R-2234/1/521/193

The total mass of the metal reinforcements for each vessel was quite significant. From the table below we can see how the mass was distributed for each vessel.

Table 2. Mass of dredges.

Msta	*Lena*	*Pechora*	*Zeia*
38,000 kg	32,290 kg	32,290 kg	30,270 kg

The difference in the mass of the reinforcement for the three identical vessels (*Msta, Lena,* and *Pechora*) is explained by the fact that the shipyard of Murmansk was incapable of completing the massive assignment of ice-strengthening four vessels in the planned eight days (beginning from 15 July 1939). In his note to Joseph Stalin [12, l. 12], Andrei Dubravin wrote that the shipyard did not provide the vessels with a dry dock. In result, only *Zeia* was furnished according to the original project. The rest of the vessels set sail in different stages of completion. Nonetheless, even the incomplete reinforcement and hull sheathing greatly enhanced the dredges. The table below compares the strength of *Lena* before and after strengthening [12, l. 89].

Table 3. Strength of vessels before and after reinforcement.

Hull section	**Maximum evenly distributed stress on each frame**	**Maximum evenly distributed stress on hull plates (tons/m^2)**	**Maximum evenly distributed stress longitudinal elements**

			(stringers) (tons/m^2)
Unprotected	2.3 tons/m^2	13.4 tons/m^2	None
Protected	7.2 tons/m^2	53.5 tons/m^2 (in the area of the wooden sheathing)	23.5 tons/m^2

As we can see, the strength of the vessel increased significantly. Dubraving calculated that this protection would suffice the ships to sail through the icy waters of the Northern Sea Route and withstand both transverse pressure exerted by the ice floes and mechanical deformation that could be caused by the hulls repeatedly beating the ice.

4. The voyage.

In late July 1939, the ships *Zeia*, *Msta*, *Pechora*, and *Lena* sailed from Murmansk [12, l. 102]. The tugs *Groza*, *Tucha*, *Poliarnik*, and *Borets* (towed) departed from Arkangel'sk. Escorted by the icebreaker *Lenin*, the caravan passed the strait of Iugorskii Shar encountering rare ice fields and called in Port Dikson. The cargo steamer *Mossovet* carried supplies of fresh water, coal, and spare parts to the riverine vessels. From Dikson, the caravan was escorted by the icebreaker *Stalin*, and then the icebreaker *Fedor Litke*. Having favorable ice conditions, the four dredges sailed without icebreaker escort from the Medvezhyi Islands in the East Siberian Sea to Providence Bay, where they called in on 29 August [12, l. 107verso]. Further, the caravan sailed along the Asian coast to the Sea of Okhotsk, arriving at Nikolaevk on 25 September. The whole voyage took just over 55 days [12, l. 102verso]. The average speed of the caravan was 5 – 5.5 kn along the Northern Sea Route. It was reduced to 2 – 4 kn in ice (in total, 500 nautical miles of the voyage were in ice-infested waters). The damage to the hulls was minimal (Fig. 3).

Fig. 3. Bow section of ice protection damaged by ice. Source: RGAVMF file R-2234/1/521/40

4. Conclusion

In the long run the expedition of 1939 was just a commonplace shipping operation on the Northern Sea Route, which by the end of the 1930's had been more or less developed for seasonal navigation. The uniqueness of the expedition was in its vessels – expensive imported machinery, which required special ice protection. While there is no direct evidence that Dubravin used the experience of earlier shipwrights in his design of the hull sheathing and interior reinforcement, it is reasonable to suggest that he grasped some of the engineering solutions from polar literature on past expeditions. If this is so, then from the aforementioned description we can derive the construction of the ice protection of polar ships, used in earlier epochs.

References

[1] Knight R J B 2013 The introduction of copper sheathing into the Royal Navy, 1779–1789 *The Mariner's Mirror* 59(3) pp. 299–309.

[2] *Dictionary of Newfoundland English* (Story G M, Kirwin W J, Widdowson J D A (eds.)) 1990 (University of Toronto Press).

[3] Nansen F 2008 *Farthest North: the epic adventure of a visionary explorer (The Norwegian Polar Expedition, 1893–1896)* (Skyhorse Publishing, Inc).

[4] Beechey F W 1832 *Narrative of a voyage to the Pacific and Beering's Strait to cooperate with polar expeditions: performed in HMS Blossom, under the command of Captain F.W. Beechey in the years 1825, 26, 27, 28* (Philadelphia: Carey & Lea).

[5] Parry W E 1821 *Journal of a voyage for the discovery of a North-West Passage from the Atlantic to the Pacific: performed in the years 1819–20 in HMS Helca and Griper, under the orders of William Edward Parry* (London: John Murray).

[6] Parry W E 1824 *Journal of a second voyage for the discovery of a North-West Passage from the Atlantic to the Pacific: performed in the years 1821–22–23, in HMS Fury and Hecla, under the orders of William Edward Parry* (London: John Murray).

[7] Battersby W and Carney P 2013 Equipping HM ships *Erebus* and *Terror*, 1845, *The International Journal for the History of Engineering & Technology*, 2013, 81:2, pp. 192–211, DOI: 10.1179/175812111X13033852943147

[8] McClintock F L 1859 *The voyage of the 'Fox' in the Arctic Seas. A narrative of the discovery and the fate of Sir John Franklin and his companions* (London: John Murray).

[9] Peel H 1894 *Polar Gleams* (London : E. Arnold).

[10] Dubravin A I 1983 *Kolymskii reis parokhoda "Stavropol'" [The Kolyma voyage of the steamer Stavropol']* (Magadan: Knizhnoe izdatel'stvo).

[11] Belov M I 1969 *Nauchnoe i khozyaystvennoe osvoenie Sovetskogo Severa 1933 – 1945 gg. [The scientific and economic development of the Soviet North, 1933 – 1945]* (Leningrad).

[12] Russian State Naval Archive (RGAVMF) (St Petersburg, Russia) file R-2234/1/42.

IOP Conf. Series: Materials Science and Engineering **467** (2019) 012016 doi:10.1088/1757-899X/467/1/012016

Prevention of porosity formation in power unit castings from aluminum alloys

G G Krushenko, O A Platonov, V P Nazarov
Reshetnev Siberian State University of Science and Technology, Krasnoyarsk, Russia

E-mail: platonov_oa@sibsau.ru

Abstract. This paper discusses the mechanism of porosity formation in castings from aluminum alloys, measures in preventing their education, and ways of blocking the negative impact on castings properties.

1. Introduction

Foundry alloys on the basis of aluminum [1] are widely applied to receiving castings of a machine-building profile [2], including case details of power units aviation [3] and the space equipment [4] on account that these alloys are characterized, high technological [5], and mechanical properties [6] which guarantees their work reliability in the difficult loaded conditions, for example, in the structure of fuel-pump units (Figs. 1, 2) LPE [7, 8].

Fig. 1. Cast detail of the turbo-pump unit [4]
(http://www.yuzhmash.com/ru/product.php?page=uzliavto)

Fig. 2. LPE main turbo-pump unit assembled [8]

2. Experimental

Generally for this purpose alloys on the basis of the Al-Si and Al-Mg system are applied with implanting their various strengthening additives [9].

IOP Conf. Series: Materials Science and Engineering **467** (2019) 012016 doi:10.1088/1757-899X/467/1/012016

However it is known that aluminum alloys during their preparation process in a varying degree dissolve hydrogen [10] which content in liquid metal is defined by the melting temperature and time modes and fusion processing the means of which interfere in hydrogen saturation [11]. Usually decontamination is made by introducing chlorine-containing compounds to the fusion, for example, chloride magnesium MnCl2 or the more effective C2Cl6 hexachlorophene. As a result in the interaction of these connections with hydrogen HCl couples which disappear from fusion are formed. There are other means, and ways of reducing hydrogen content in fusion (e.g. a direct chlorine or nitrogen purge, current processing, ultrasound and etc.). The effective way to increase the castings density without contaminating them with the products which form during the interaction of decontaminators with hydrogen and with the melting units lining, consists in creating directed metal crystallization in the shape from the lower layers of casting to the upper, as a result of cooling the chill mould bottom [12].

The hydrogen dissolved in liquid metal in the process of the castings crystallization received from aluminum alloys with a narrow crystallization interval (temperature from the beginning to the end of crystallization from the line of a liquid to the line of a solidus to which the Al-Si system alloys belong to (Fig. 3) [13] in the range of the silicon content from 8,0 to 11,5% (for example, the steel grades Ak9ch, Ak7ch, etc.) forms scattered porosity in the volume.

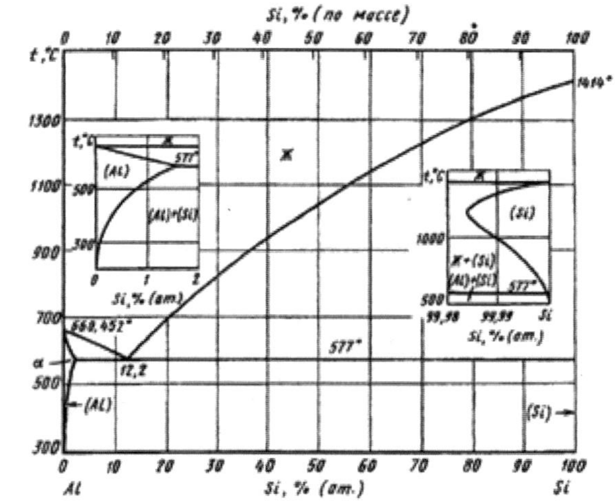

Fig. 3. States of Al-Si chart [13]

The assessment of castings porosity (that is the quantity and the size of the pores) is made in conditional scores (Table 1, Fig. 4) while studying the x-ray films exhibited by means of x-raying during the imposition of templates (flat samples) cut out from a detail representative. At the same time shrinkable friability or central porosity is excluded. It is possible to use porosity degree standards independent to the alloy brand. Generally castings with porosity no more than 1 and 2 scores are allowed to production, but on the details working in the high pressure conditions porosity isn't allowed at all. It has been established that the density of castings from aluminum alloys minerals reflects the presence of discontinuities, and their number and size influence the level of the cast parts mechanical properties. The pores at aluminum castings reduce metal density which leads to decrease in their mechanical properties [14]. For example, in the Ak9ch alloy (8,0-10,5% of Si; 0,17-0,30% of Mg; 0,20-0,50% of Mg; rem. - Al) we determined [15, 16] that at the density of castings (determined by the method of hydrostatic weighing) ρ = 2564 kg/м^3 the temporary resistance of $\sigma_в$ = 235 MPas, relative elongation of δ = 5,0% whereas at ρ = 2575 kg/m^3, $\sigma_в$ has increased to 260 MPas (by 10,4%), and δ - to 5,6% (for 12,0%), and at ρ = 2658 kg/m^3 $\sigma_в$ has increased to 275 MPas (by 17,0%), and δ– to 8,2% (gaining 64,0%).

Table 1 [1]

Porosity, score	Pore diameter, mm	The number of pores per 1 cm^2, a piece
1	to 0,1	to 5

Porosity, score	Pore diameter, mm	The number of pores per 1 cm^2, a piece
2	to 0,1	to 8
	to 0,2	to 2
3	to 0,3	to 12
	to 0,5	to 3
4	to 0,5	to 14
	to 1,0	to 6
5	to 0,5	to 15
	to 1,0	to 8
	> 1,0	to 2

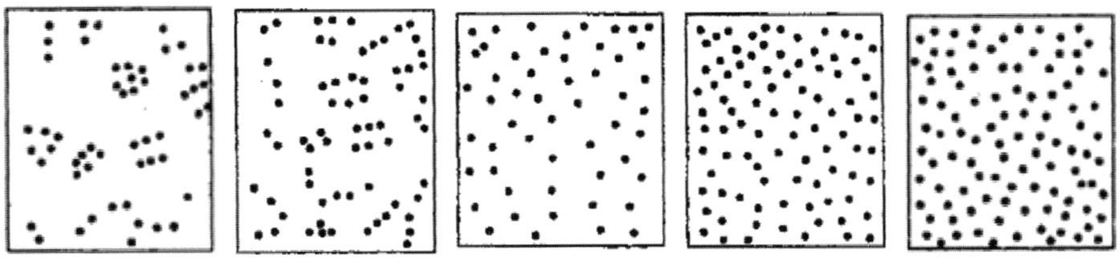

score 1 score 2 score 3 score 4 score 5

Fig. 4. Porosity by the VIAM scale

When casting details from alloys with a wide interval of crystallization to which alloys of the Al-Mg system belong (Fig. 5) [13] in the content range of magnesium from 6,0 to 10,5% (for example, AL8, Amg10ch, etc.), on account of metal crystallization on the so-called "volume" mechanism, castings are struck with interdendritic shrinkable porosity (Fig. 6) [17] which reduces their durability and hydraulic soundness (tightness). This defect usually arises at a slow speed of alloy crystallization, for example, when casting to sandy-argillaceous forms. At the same time in the shrinkable emptiness hydrogen dissolved in fusion is formed.

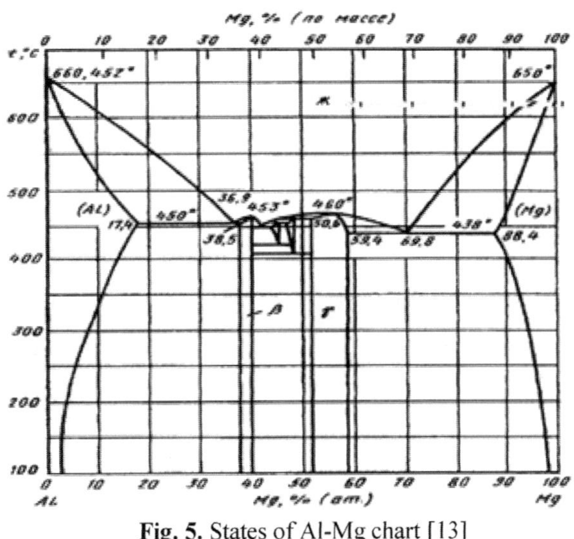

Fig. 5. States of Al-Mg chart [13]

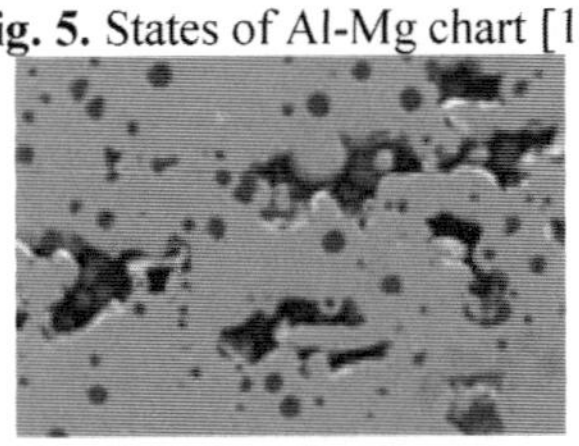

Fig. 6. Shrinkable porosity [http://www.lityo.com.ua/li/s_41.html]

On the basis of the alloys under pressure crystallization theory offered by the famous metallurgist and academician A.A. Bochvar who worked with professor A.G. Spassky at the Moscow Institute of Non-Ferrous Metals and Gold technology of autoclave casting aviation engine details from aluminum alloys was developed and introduced to production. The collected form was established in metal content in the cover of which there was an opening that corresponded to a pouring cup in a casting mould. After filling the metal in a form the cover in the metal content was quickly closed, and supplied with compressed air under 4 ... 5 atm pressure. (1 atm = 0,10133 MPas = 1,0333 kgfs/cm^2) [18]. The pressure on the liquid metal during the crystallization process strengthens the supply with the help of the hardened and interdendritic emptiness fusion. As a result dense castings mechanical properties and hydraulic soundness (tightness) of which are in accordance with the technical documentation are formed.

3. Conclusion

There is a method of increasing density in the already cast details which is reached by impregnating the shrinkable emptiness with various substances and their subsequent hardening [19, 20]. As an impregnating substance a Bakelite varnish is often applied. Impregnation is carried out in the autoclave under pressure with the subsequent washing of the impregnated details with ethanol and drying.

References

[1] GOST 1583-93 Foundry aluminum alloys.
[2] Zakharov A M 1980 *Industrial alloys of non-ferrous metals. Phase structure and structural components* (Moscow).
[3] Starke E A (Jr) and Staley J T 1996 Application of modern aluminum alloys to aircraft *Progress in Aerospace Sciences* Vol 32(2–3) pp. 131–172.
[4] Krushenko G G and Kashubsky A N 2010 Application of an X-ray analysis for identifying and welding-up defects in critical duty castings from aluminum-silicon alloys *Repair, restoration, modernization* No 8 pp. 47–49.
[5] Korol'kov A M 1967 *Foundry properties of metals and alloys* (2nd ed.) (Moscow).
[6] Stroganov G B 1985 *High-strength foundry aluminum alloys* (Moscow).
[7] Moiseyev V A et al 2008 *Production technology of liquid rocket engines* (Moscow: MSTU).
[8] Makhutov N A et al *2011 Durability and ZhRD resource* (Moscow).
[9] Altman M B and Stromskaya N P 1984 *Increase in properties of standard foundry aluminum alloys* (Moscow).
[10] Chernega D F and Byalik O M 1972 *Hydrogen in foundry aluminum alloys* (Kiev: Tekhnika).
[11] Postnikov N S, Mel'nikov A V, Lebedev V M 1971 *Melting of aluminum alloys* (Moscow).
[12] Krushenko G G 2008 Automatic control of casting metal form cooling *Mechanical engineering and automation problems* No. 4 pp. 105–106.
[13] Vol A E 1959 *The construction and properties of double metal systems* (Moscow).
[14] Irfan M A et al 2012 Porosity reduction and mechanical properties in die engine blocks *Materials science and engineering* Vol. 535 pp. 108–114.
[15] Krushenko G G and Vasilenko Z A 1988 Density and mechanical properties of the alpaxes which are thermally processed in liquid state *Fusions* 2(6) pp. 67–69.
[16] Vasilenko Z A et al 1992 The influence of condition in the form of cast test samples on the aluminum alloys mechanical properties *Problems of durability* No. 1 pp. 80–82.
[17] Gunasegaram D R, Farnsworth D J, Nguyen T T 2009 Identification of critical factors affecting shrinkage porosity in permanent mold casting using numerical simulations based on design experiments *Journal of Materials Processing Technology* Vol. 209 pp. 1209–1219.
[18] Spassky A G 1950 *The bases of foundry production* (Moscow).
[19] Impregnation improves casting quality 1953 *Vacuum* 3(1) p. 94.
[20] Jolly M 2003 Castings *Comprehensive structural integrity* Vol. 1. pp. 377–466.